Stellar Explosion near the Trapezium in Orion
Credit: ALMA (ESO/NAOJ/NRAO), J. Bally/H. Drass et al.

Beyer's *Star Guide* is the best work of its kind that I have ever encountered, a delightful blend of practical instruction and the history of astronomy and its legends." – Nelson Bryant, *The New York Times*

"What most impresses me about this work is the way it makes those remote, scintillating pinpoints of light come alive as astronomical bodies. A refreshingly different introductory sky guide, truly approachable, from which the layperson can learn some real astronomy" – George Lovi, *Sky and Telescope*

"A well-thought out, non-complex approach to presenting concepts. The charts are not cluttered with labels and faint stars so that the patterns of bright stars are not interfered with. *The Star Guide* is a good investment for the library of a beginning stargazer and will certainly become one of the most dog-eared books in such a library." – Robert A. Garfinkle, *Astronomy Magazine*

"A full reading of all section texts will give a concise introduction to astronomy as a science, but the book really isn't meant to be read straight through. It is intended that the reader pick it up on a cloudless evening, turn to the current section and take it outside to find the stars." – Robert J. Chambers, *Los Angeles Times*

"A fascinating approach for beginning and amateur astronomers, the book introduces over 100 of the brightest stars. After a few weeks of observations, the bright stars will serve as reference points to guide the reader. Well illustrated." – *Science News*

"*The Star Guide* provides the backyard naturalist with a simple but effective method for learning the 105 brightest stars." – *Science Books and Films*, American Association for the Advancement of Science

"This is a guide for the serious amateur who will find it genuinely useful for field study and reference." Lee Pennock Huntington, *The Sunday Rutland Herald*

"A brief introduction to the history of astronomy enhances this practical, exciting guidebook." – *Booklist*, American Library Association

"*The Star Guide* is a work of long and hard research. Each constellation and its important stars are discussed in detail. Their lore, spectra, luminosities, distances, and other information are treated in a lucid and lively fashion. I recommend *The Star Guide* to the library of any amateur astronomer." – John Marshall, *Event Horizon*, The Journal of the Amateur Astronomers Association of New York

"A boon to the budding astronomer" – Mark Orwoll, *Aviation Week & Space Technology*

"One learns to locate the stars again and again. A hand-sized course for beginners ranging from students to the general public" – The Research Libraries, New York Public Library

"At last, a guide that tells you how to locate stars in a manner so simple that even a city dweller who rarely looks upward can understand it." – *Chicago Southtown Economist*

"*The Star Guide* is truly an outstanding and most useful exposition of the visible sky." – Fred Hess (namesake for Main-belt Asteroid 2844 Hess), *Recent Publications in*

Natural History, Department of Library Services, American Museum of Natural History

"A Delightfully Different Observer's Companion, I am sorry to see that this work is currently out-of-print (c. 1998) as it is, as the title states, a rather unique guide to exploring the night sky. Unlike most others of its kind, Beyer's work truly places its emphasis on the stars themselves and their seasonal position as opposed to their constellations (although this information is certainly included). It truly is a delightful astronomical companion for the more serious amateur observer, with many interesting bits of lore and science about individual stars that I've never encountered elsewhere." – Cato, Reviewed in the United States

"This is an excellent guide while viewing stars. Very readable star guide. Includes locations, meaning, lore, history, and descriptions of the brightest stars in the night sky. Great to have with you while looking through your scope!" Reviewed in the United States

"I didn't give much thought to constellations until an evening course at the Hayden Planetarium given by Steven L. Beyer, author of The Star Guide: A Unique System for Identifying the Brightest Stars in the Sky. (That was our "text" and remains a well-worn treasure!) Over four nights, our little group was ushered into the planetarium, which we had all to ourselves!!! The giant dome was set for either Winter, Spring, Summer, or Fall. It was a killer on the neck! But it was wonderful to be gazing up listening to this expert's lecture on that season's night sky, including all sort of legends and lore."

– Paula, Reader's Comment, *This Week's Amazing Sky, The Old Farmer's Almanac*

Cone Nebula, Christmas Tree Nebula (inverted), and Fox Fur Nebula Credit: ESO

THE STAR GUIDE

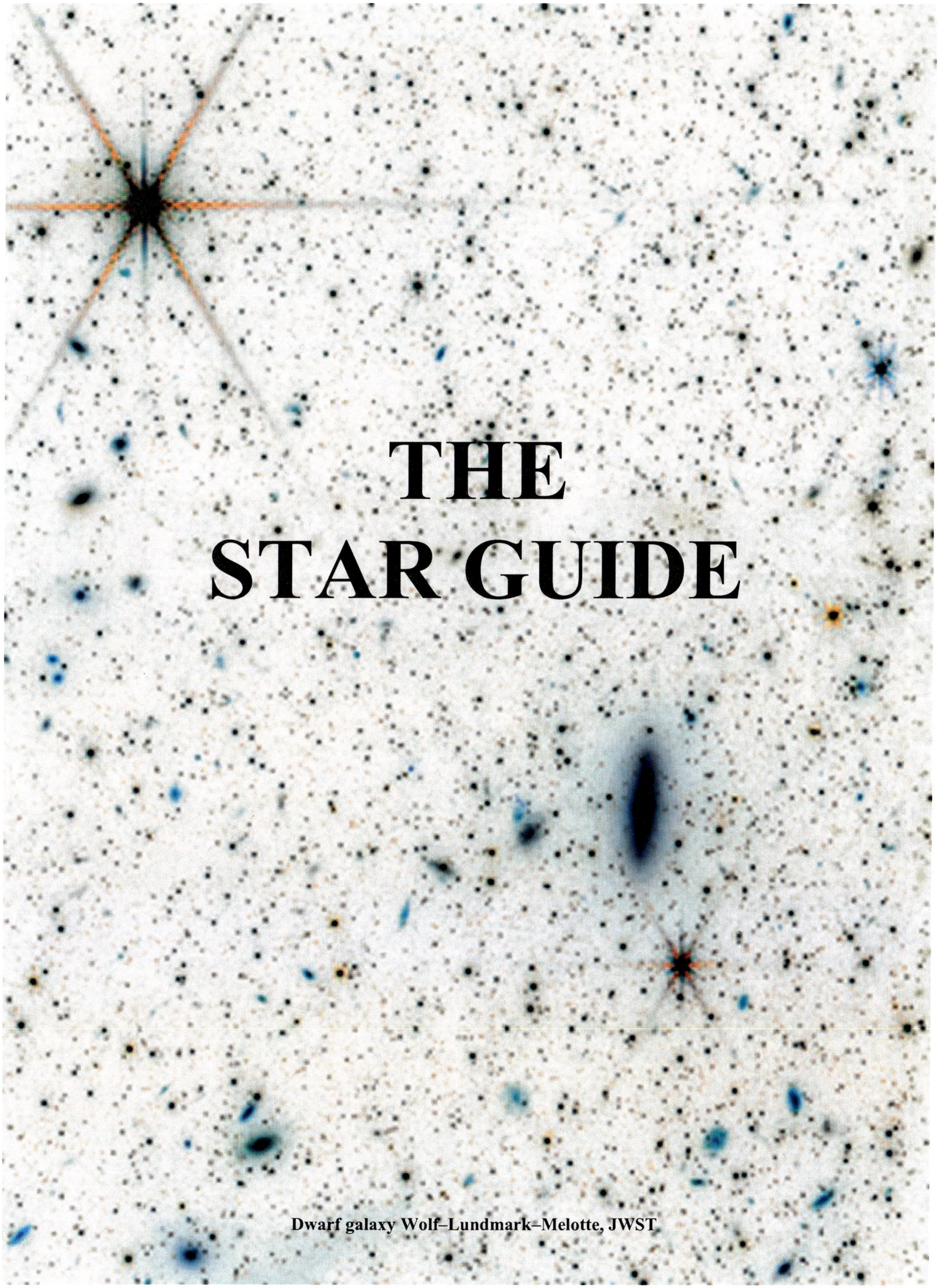

Dwarf galaxy Wolf–Lundmark–Melotte, JWST

Star Birth in Orion - Herbig-Haro Object 124 – Polar Jets and Nebula
Credit: ESA/Hubble & NASA

THE STAR GUIDE

A Unique System for Identifying the Brightest Stars in the Night Sky
Revised – CELESTIAL SPORT 8.511 EDITION 3 Volume Four

Guide to the Universe in Four Volumes:

1 – Introduction, Rising Stars of Winter
2 – Spring Risings, The Universe of Stars
3 – Summer Arrivals, Knowing About Stars
4 – Rising Stars of Autumn, Continuations

M92 globular star cluster in Hercules, left of center view - JWST
Credit: NASA, ESA, CSA

THE STAR GUIDE

A Unique System for Identifying the Brightest Stars in the Night Sky Revised,
CELESTIAL SPORT 8.511 Edition 3

Volume Four – Rising Stars of Autumn, Continuations
The 26 Essentials

BY

Steven Larsen Beyer   Andrea Elgin Beyer

Galaxy field with three extremely distant (circled) examples in the constellation Fornax
They appear here as they were, just 830 million years after the Big Bang.
In the center ring is a quasar whose relative brilliance dominates all stars in its surrounding galaxy.
Credit: JWST/ NASA, ESA, CSA

Correspondences regarding this work should be sent to Beyer Publications: thestarguiderevised@earthlink.net

Revised, Celestial Sport 8.511 Edition 3, Volume Four (pbk.) ISBN 978-1-7360454-6-6
Revised, Celestial Sport Edition (2), Volume Three (pbk.) ISBN 978-1-7360454-1-1
Revised, Celestial Sport Edition (2) ISBN 978-1-7360454-0-4
First Edition ISBN 0-316-09267-3
First Edition (pbk.) ISBN 0-316-09268-1

THE STAR GUIDE: A Unique System for Identifying the Brightest Stars in the Night Sky
Revised CELESTIAL SPORT 8.511 EDITION 3 Volume Four - Rising Stars of Autumn, Continuations

Cover Images: JWST First Deep Field, Hubble Ultra Deep Field, & The Crab Nebula (5λ) / NASA, ESA, G. Dubner (IAFE, CONICET-University of Buenos Aires) et al.; A. Loll et al.; T. Temim et al.; F. Seward et al.; VLA/NRAO/AUI/NSF; Chandra/CXC; Spitzer/JPL-Caltech; XMM-Newton/ESA; and Hubble/STScI.

Sky Screen, Seasonal, and Celestial Sport Star Charts: by the Authors
We wish to thank Eric Elgin Beyer for Edits and Comments.
Adobe Stock Images (asi)   Shutter Stock Images (ssi)   Other images as indicated

Star Hemisphere Circles - Stellarium
Credit: IAU S&T celestial maps:
Authors: International Astronomical Union and Sky & Telescope magazine (Roger Sinnott & Rick Fienberg)
Source: International Astronomical Union (IAU) / constellations

Dedicated to those who set us on course for our stars to align:

Sophie (Sunny) Larsen Beyer, Melba Thompson Elgin, Svend Aage Beyer, Frank Roy (Pete) Elgin, Mary M. Musick, Principal Kings Park Elementary, Fairfax County, VA, Marcy Pedas Sigler

The Southern Ring Nebula, Constellation Vela, 2,000 light years from Earth – JWST Near-Infrared Camera
Credit: NASA, ESA, CSA, STScI

He who has not contemplated the mind of nature
which is said to exist in the stars
is not able to give a reason
of such things as have a reason.
Plato, *Laws XII*

There are more things in heaven and Earth, Horatio,
than are dreamt of in our philosophy.
Shakespeare, *Hamlet – Prince of Denmark*

“Snake” in the Galactic Plane - Sagittarius. Spitzer Space Telescope Credit: NASA/JPL-Caltech

Front Gallery –
Page
ii The Atacama Large Millimeter/submillimeter Array (ALMA) radio telescope shows a stellar explosion as new stars are being formed in the Orion Nebula near the four Trapezium stars – seen at lower left in this composite. The view also includes images from the Gemini South and ESO Very Large Telescope Credit: ALMA (ESO/NAOJ/NRAO), J. Bally/H. Drass et al.

iv NGC2264 The Cone Nebula, about 2,400 light years from us, is at bottom of this image - under the upside-down Christmas Tree Nebula which fans outward and up with star S Monocerotis representing the tree trunk at top of the picture. The Fox Fur Nebula is at the top right corner, all in the constellation Monoceros. Credit: European Southern Observatory

v Portion of Dwarf galaxy Wolf–Lundmark–Melotte (WLM), a barred, irregular type, 3 million light years distant from us in the constellation Cetus at coordinates 00h 02m −15° 28′ (negative image) NIRCam - Near-InfraRed Camera Credit: NASA, ESA, CSA, STScI, and K. McQuinn (Rutgers University), A. Pagan (STScI).

vi Herbig-Haro Object 124 in Orion Hubble Space Telescope Image w. Jet and nebula
Credit: ESA/Hubble & NASA, D. Padgett (GSFC), T. Megeath (University of Toledo), and B. Reipurth (University of Hawaii)

viii M92, globular star cluster in Hercules, left side view at 27,000 light years distance, 17h 17m +43° 08′ NIRCam
Credit: NASA, ESA, CSA, A. Pagan (STScI)

ix M92 right side of the JWST camera's occulting strip, negative view Credit: NASA, ESA, CSA, Alyssa Pagan (STScI)

ix Portion of ASPIRE J0305-3150 (RA: 03h 05m dec: -31° 50') image includes circled galaxies at high redshift in constellation Fornax JWST, NIRCam (Near-Infrared Camera) Credits: NASA, ESA, CSA, Feige Wang (University of Arizona) Image Processing, Joseph DePasquale (STScI)

xii The Southern Ring Nebula, NGC3132, the Eight-Burst Nebula, Constellation Vela, 2,000 light years, Webb's Near-Infrared Camera (NIRCam) Credit: NASA, ESA, CSA, STScI

xiv "Snake" in the Galactic Plane - infrared image from NASA's Spitzer Space Telescope The snake nebulosity is located about 11,000 light-years away in the constellation Sagittarius. credit: NASA/JPL-Caltech

xvi Herbig-Haro Objects 46/47 - Nascent Star Nebulae in the Constellation Vela, 1,470 light years from Earth – James Webb Space Telescope Credit: NASA, ESA, CSA, J. DePasquale (STScI)

xviii Barnard Loop, Orion - bright, 20,000K emission nebulae
https://commons.wikimedia.org/wiki/File:BarnardLoopHunterWilson.jpg – Credit: Hunter Wilson

xx NGC 1433 Seyfert galaxy – with bright active core containing a super massive black hole, distance 46 million light years from Earth in the constellation Horologium. James Webb Space Telescope, Mid-Infrared Instrument (MIRI) image, camera and a spectrograph - mid to long infrared radiation from 5 to 28 microns. Overall brightness and lack of dust suggests recent merger with another galaxy, causing an intense episode of dust-clearing star formation activity. Credit: NASA, ESA, CSA, Janice Lee (NSF's NOIRLab) Image Processing Alyssa Pagan (STScI)

2 Messier 35 Open Star Cluster - 3,000 light years from Earth (diffuse upper right) and more concentrated open cluster NGC 2158 c. 11,000 light years to the lower left – in the constellation Gemini – Credit: 2MASS (Two Micron All-Sky Survey)

Herbig-Haro Objects 46/47 - Nascent Star Nebulae in the Constellation Vela – James Webb Space Telescope

Contents

Barnard Loop, the Orion Nebula M42, and M43 – Credit Hunter Wilson

Preface – Guiding Stars

Stars appear as beautiful sparkling points of light, and their lofty presence inspires us to learn more about them. The night sky is one of nature's visual treasures as well as a manifestation of our ultimate origin and destiny. Our hope is that using *The Star Guide* will increase your appreciation of the sky's magnificence.

We are progeny of stars in the sense that most of the chemical elements in our bodies were forged during the evolution of ancient stars. Matter from those stars later coalesced to form Earth, and for approximately 4.6 billion years the Sun, our nearest star, has provided energy that continues to animate life on this planet. During that time it is likely various other forms of celestial energy, including ultraviolet light and cosmic rays, also helped alter the course of biological evolution by affecting genetic material in plants and animals.

During the Devonian Period of Earth's history, about 400 million years ago, some life forms began a migration from sea to land. Today we find ourselves on the threshold of a far wider ocean, whose ultimate challenges and rewards can just be imagined. As we gaze at stars from a sidewalk or backyard, we look towards a realm of space and time that beckons us to life's next arena in its majestic passage through the ages.

Humans now dominate other forms of life on Earth. We challenge each other and the deepest mysteries of nature. At this stage we may indulge in the luxury of savoring our natural world, looking at stars, and dreaming of our origins. As you watch stars or listen to the thunder of ocean waves, you can sense our distant heritage.

It is said Beethoven (1770-1827) was inspired to compose the second movement (at c. 10:50 min.) of his String Quartet No. 8, Opus 59 No. 2 in E Minor (1806), while gazing at the night sky and contemplating music of the spheres on a summer evening from the town of Baden near Vienna. He noted the piece is to be played with much feeling. This music has been described as being evocative of emotions we may experience while embracing the night sky's glory. Vincent van Gogh (1853-1890) was also inspired by the stars, as may be seen in his *The Starry Night* and *Starry Night Over the Rhone.*

In addition to sublime visual aspects of the stars, astronomers have learned much about their physical characteristics. *The Star Guide* will introduce you to many of the brightest, so they will become familiar and can evoke fond memories of the times when you learned their names and how to identify them. Ability to recognize individual stars is a skill which is a pleasure to enjoy alone or to share with friends.

Essayist and historian Thomas Carlyle (1795-1881) wrote: "Why did no one teach me the constellations when I was a child?" This lament need not apply to you. Use *The Star Guide* to make friends with the stars and in this process enhance your sense of place in the universe.

The following bright star calendar of Autumn months follows the plan of our Winter, Springtime, and Summer volumes. Dates highlight when each featured star reaches about a third of the way up from the horizon around 10 p.m. (until we "fall back" to Standard Time on the first Sunday in November, when nine p.m. is suggested). Essential stars are then hopefully clear of low sky mists, ground obstructions, and it's not too close to bedtime!

After each subsequent week, stars reach Sky Screen introductory positions a half hour earlier as prime evening viewing seasons continue. Stars rise about four minutes earlier every day, and Earth's orbital progression appears to move them gradually westward until observing seasons end with each star approaching the western horizon at sunset, their glow becoming lost in a flood of evening twilight. Except for circumpolar objects seen in the far northern part of the sky which perpetually dodge the horizon as Earth rotates and revolves, most featured stars are invisible for periods of time up to several months; then each is seen again low in the east during their heliacal risings just before morning twilight becomes too bright.

Southerly stars have rather brief prime evening seasons for viewers at mid-northern latitudes. For example, Fomalhaut at the "mouth" of Piscis Austrinus – the Southern Fish, is seen quite easily during evenings of just October and November while stars further north on the celestial sphere offer progressively longer prime seasons. Brilliant Capella, brightest in Auriga – our essential star for November 2nd by comparison has a prime evening season lasting through autumn and winter into the following April, and may be seen at some time during the night all year!

Sky Screen fields of view encompass the wide area most of us see when looking in a particular direction. They show each featured star at the middle of a sky region with one or more other essential stars.

Celestial sport encourages star identification based on noticing and remembering concise patterns such as triangles, doglegs, and arcs of featured essential stars – all visual magnitude 2.9 or brighter. Auxiliary charts with fainter stars to magnitude six are also provided for your use at darker sky locations.

Traditional creative representations of traditional constellation figure outlines generally show far more faint stars than practical to see from places where most people live these days.

Even when looking skyward from a truly dark site, where many hundreds of stars are visible, experienced sky watchers usually seek first the brightest essential stars, then "hop" to their less vivid neighbors for orientation. Stellar geometric patterns we present in our Sky Screens generally include members from several adjacent constellations, not relying just on elements of a featured star's traditional constellation outline.

NGC 1433 – Seyfert galaxy with extremely active core surrounding a supermassive black hole - JWST

October – Rising Bright Star Introductions

Earth at Mid Orbit

(Prime Evening Season)

2 **Mirfak** – Perseus October – March

4 **Algol** – Perseus October – March

17 **Epsilon Persei** – Perseus October – March

21 **Fomalhaut** – Piscis Austrinus Oct. – November

22 **Diphda** – Cetus November – January

24 **Alcyone** – Taurus November – March

27 **Menkar** – Cetus November – January

Earth at Mid-Orbit

Like a bouncing ball that reached its peak and now is on the way down, Earth has been moving closer to the Sun since early July's aphelion. During the first week in October our planet again is at its average annual distance from the nearest star. Separation continues to decrease until the beginning of January when Earth achieves perihelion.

Average Earth to Sun distance is reached around October 5th, about 92.9 million miles / 149.6 million kilometers. In early October we're about one and a half million miles / two and a half million kilometers closer to the Sun than we were during early July. We'll keep getting slightly nearer until minimum separaion is reached, at the "bottom" of Earth's bounce during January's perihelion, about 91.4 million miles / 147.1 million kilometers.

Representing Earth's average orbital position every year in early October, a poppy seed size Earth model placed on a line toward first base would be 26 feet, two inches, from a baseball sized Sun model on home plate. That's a decrease of five inches from the scaled Earth-Sun aphelion distance at the start of July, almost 30% of the way from home to first base.

Scaled soccer ball/Sun to pinhead/Earth model distance at July's aphelion decreases by about 40 centimeters to 23.6 meters from the field's center spot by early October. That's about half the distance from center spot to the front line of a goal box.

Earth's average annual orbital speed on the mid-range dates of early October and April are about 66,600 mph /107,200 km/hour.

Messier 35 at upper right and NGC 2158 to lower left - Open Star Clusters in Gemini – Credit 2MASS

Mirfak October 2

North-Northeast Northeast East-Northeast

Perseus

October – March (prime evening season)

MERE-fak

1.8 mag. white supergiant 510 ly

Anchor Point Star, F5 Ib

****

Imagine the Sun being the size of a baseball. By comparison, Earth would be 1/32 of an inch across, slightly smaller than a pinhead. The model Earth would to scale be placed, 20 feet, one-third the distance, between the baseball/Sun model set on home plate and first base.

At the same scale, Mirfak would be about 16.5 feet wide, 91% the width of a pitcher's mound, and located 158,900 miles from our ball/Sun model – 67% of the actual distance between Earth and the Moon. Compared with a soccer ball sized Sun model on center spot of a pitch, Earth would be two millimeters wide – twice the size of a pinhead.

Using this metric scale, the model of our planet would be halfway between the Sun/ball and one of the goals.

Proportionately, a model of supergiant Mirfak would be huge, about 15 meters in diameter – 82% as wide as the soccer field's center circle. At its appropriately scaled distance, the Mirfak model would have to be moved 762,500 kilometers from the soccer ball/Sun – that's about twice the actual distance between Earth and the Moon.

LOCATION:

Second-magnitude Mirfak is the brightest star in constellation Perseus, the Hero. It is 17 degrees east of Almach, tagging on to the curving line of second magnitude stars in Andromeda which stretches eastward from Alpheratz and includes Mirach.

MEANING:

The name Mirfak originated with an early Arabian description of stars in Perseus as the "Elbow of the Pleiades", adapted from a word for elbow.

VISUAL ASPECTS:

White supergiant star Mirfak lies at a distance of about 510 light years from Earth. It's about 4,000 times more luminous than the Sun and has surface temperature of approximately 6,400 K. With an apparent magnitude of 1.8, Mirfak highlights delicate tracery of star chains seen in Perseus.

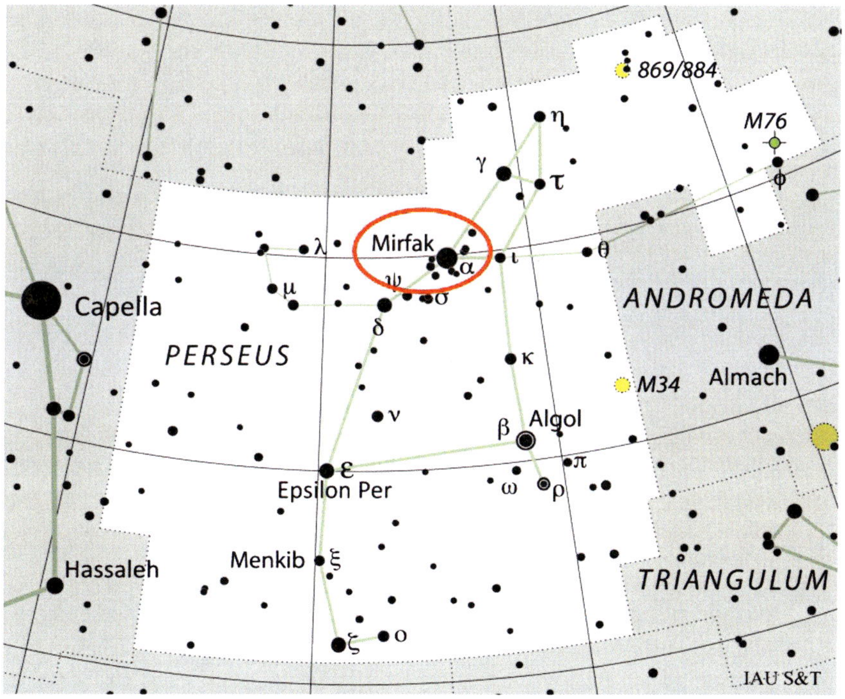

PHYSICAL NATURE:

Mirfak is another of the Anchor Point stars which define spectral and luminosity categories in the MKK Atlas. A benchmark for all type F5 luminosity class Ib supergiants, Mirfak is vast, eight and a half times the Sun's mass and 68 times the diameter of "our" star. Its volume is about 314,000 times larger than the Sun.

MIRFAK - ADDITIONAL DATA:

Catalogue Designations: Alpha Persei, α Per, 33 Persei, HD 20902, HIP 15863, HR 1017, SAO 38787

****

Parallax: 6.4 mas
Distance: 156 parsecs, 510 light years

****

Apparent magnitude: 1.8
Absolute magnitude: -4.2
Luminosity: 4,000 solar
Luminosity class: supergiant Ib

****

Apparent color blend: white
Peak color: white-yellow
Color index (B-V): 0.48
Spectral type: F5
Surface temperature: 6,400 K

****

Diameter: 68 solar

Mass: 8.5 solar

****

Coordinates: (J2000.0) 03h24m +49°52m / 51.08 +49.86
Transit at 9 p.m. December 28
Proper motion: arcseconds/year +24 -26
Radial velocity: -2 km/sec.

****

NEARBY FEATURES:

Alpha Persei Star Cluster (OB2) – Mirfak is situated amidst a splendid patch of sky rich in blue-white stars. Observers using binoculars will appreciate this highly attractive group, also known as Melotte 20 or Collinder 39, centered just south of Mirfak. Proper-motion studies have identified over 100 members of the cluster, which features young, luminous, massive spectral types O and B stars.

The faint luminosity surrounding stars in this assembly group is a remnant of an original cloud of gas from which the stars coalesced. Although most of that birthing nebula disappeared, not enough time has passed for bright cluster members to have completed their life cycle and become supernovae, or to have physically migrated far from each other. This concentrated array of luminosity featuring an OB2 association of spectral types 0 and B blue-white stars is highly attractive.

Mirfak, highlighting the lovely Alpha Persei Cluster – Credit: Martin Gembec

Center of the Gould Belt – Mirfak also marks the center of a vast elliptical ring of gas, embryonic stars, and bright massive young stars that encompasses our entire sky, skewed at an angle of about 20 degrees to the Band of the Milky Way's equatorial plane. Known as the Gould Belt array it's a huge rather flat set of objects with actual overall dimensions of about 700 by 1,000 parsecs. The Gould Belt includes star forming zones such as the molecular clouds in Ursa Minor around Polaris, Orion Nebula, Perseus OB2 Association, Serpens-Aquila Rift, and the Scorpius-Centaurus Association . The Belt crosses constellations, large and small, including Crux, Carina, Vela, Canis Major, Taurus, Auriga, Orion, and Perseus. The Belt's center appears to be about 300 light years from us in the region of Perseus.

John Herschel , astronomer son of William Herschel – noted in 1847 while making extensive observations from the Cape of Good Hope, that a group of particularly bright stars including those in Orion, the Southern Cross, and in the area of Scorpius and Centaurus, are arrayed along a great circle around Earth, angled from the plane of the Milky Way's Band. This assembly was further described by American Astronomer Benjamin Gould (1824-1896) in 1879. His involvement resulted in the term Gould Belt being applied to this grand celestial array.

By the 1960's a suspected association of Gould Belt bright stars with an expanding ring of gas in the Sun's neighborhood was described by Per-Olof Lindblad. A model proposed by Carlos A. Olano of the Argentine Institute of Radio Astronomy suggested this feature comprises an elliptical region about 1,400 by 2,400 light years in dimension, centered about 500 light years from the Sun in the direction of Mirfak. A total mass of one million Suns and age of about 30 million years was estimated. It's no accident so many notable stars are evident in that zone because the Gould Belt includes many molecular clouds having star forming potential, such as those seen in the Perseus OB 2 Association and the Polaris Flare.

A Research endeavor called the Gould Belt Survey includes research by astronomers from a variety of institutions. Its aim is to study major star forming regions within 500 light years of Earth. Telescopes associated with the survey include The Spitzer Space Telescope launched in 2003 for studies in the infrared range from three to 180 microns, the James Clerk Maxwell Telescope on Mauna Kea, Hawaii with a 15 meter wide dish antenna for submillimeter wavelength study in the far-infrared to microwave range, and the Herschel Space Observatory active from 2009 to 2013 providing observations in the far-infrared and submillimeter wavelengths range of 55 to 672 micrometers (microns).

Perseus, *the Hero*

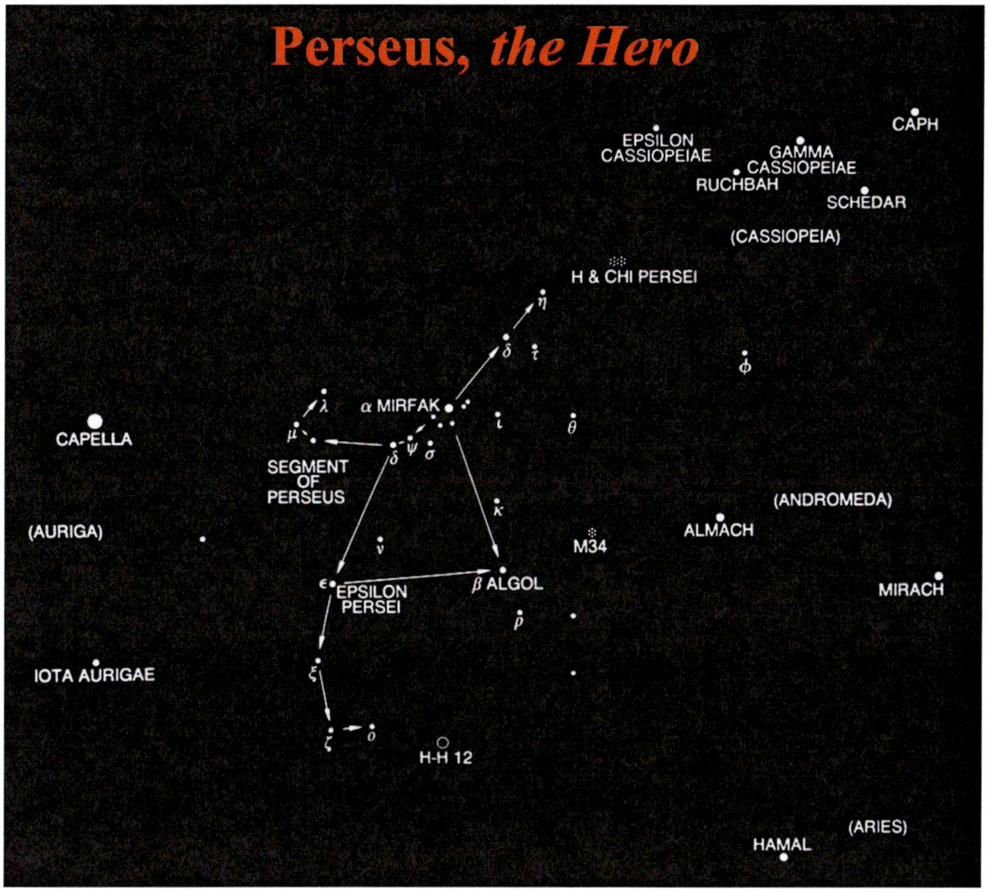

PURR-see-us, PER-syoos
Per – Persei, *PER-see-eye*
>>>>>>>>>>>>>>>>>>>>
Brightest Stars
October 2 Mirfak 1.8 white supergiant 510 ly
October 4 Algol 2.2 blue-white main sequence 90 ly
October 17 Epsilon Persei 2.9 blue-white main sequence 640 ly
>>>>>>>>>>>>>>>>>>>>
EVENING SEASON:
October - March
SKY TRACK:

Perseus follows a path across the sky from northeast to northwest. It's at the zenith when crossing the meridian, seen from mid-northern latitudes.

LORE:

A classic story involves "Superhero" stars of Perseus, one in an array of constellations seen autumn evenings and known as the "Royal Family of the Sky". The clan also includes Andromeda, Cepheus, and Cassiopeia; their legends entwined with those of Pegasus and Cetus.

Perseus is said to have been born from the union of Zeus and a princess of Argos named Danae. His grandfather, King Acrisius, had once been told by the Delphic Oracle that his daughter's son would kill him. As a precaution, the king set Danae and baby Perseus adrift in a wooden chest, hoping the sea would carry them away forever. However, the next day the chest was blown ashore on an island, where it was found by a kind fisherman named Dictys. He and his wife took Danae and Perseus into their home, where they lived happily for many years.

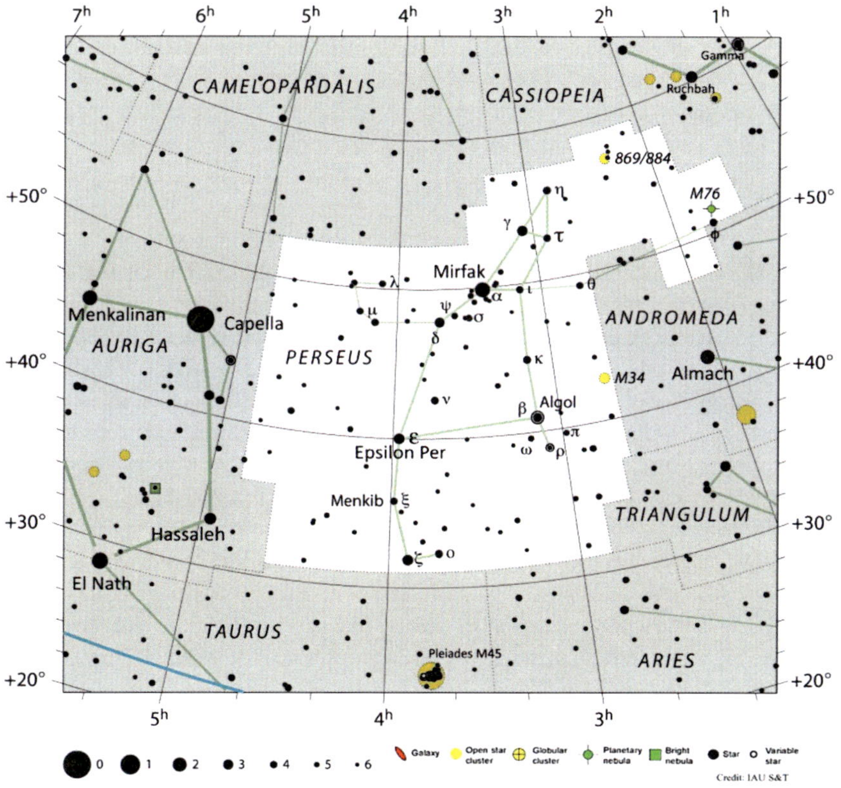

Perseus grew into manhood learning skills of a fisherman, and his still beautiful mother had become a source of attraction for the island's devious king Polydectes, brother of Dictys.

Polydectes wanted to have Danae as his wife but saw Perseus as a source of distraction and trouble. To be rid of Perseus, the king let it be known that above all else, he desired the head of the extremely dangerous Medusa as a wedding gift for Danae and him.

Proud Perseus had nothing else to offer and bravely accepted this nearly impossible challenge. Medusa was one of three hideous creatures known as Gorgons, who had body scales and wings, as well as tangles of snakes for hair. To make matters worse, anyone who looked at them would instantly be transformed into stone. Perseus took the first

ship to Greece, where he was comforted with news provided by the Talking Oaks of Dodona that he was under the protection of the gods.

Hermes came to Perseus and presented him with a magic sword, which could be neither broken nor bent. The goddess Athena appeared and gave Perseus her polished bronze shield, which he could use to view his prey indirectly and safely. Hermes then traveled with Perseus to the Northland where they tricked three women, known as the Gray Sisters, into telling them where to find the nymphs of the North. The Northern nymphs provided a great feast for Perseus and gave him a cap with the power of invisibility, a pair of winged sandals, and a leather sack. Such specialized accouterments remind us of remarkable gear associated with modern Superheroes.

Fully prepared, Perseus flew to the distant land of the Gorgons and, harmlessly viewing their reflections in Athena's mirror like shield, found them asleep. Athena and Hermes were at his side as they pointed out Medusa, the only mortal Gorgon. Guided by the reflection, Perseus descended, cut off Medusa's head and placed it in his sack. The surviving Gorgons awoke, but Perseus put on his cap of invisibility and escaped. As the Hero departed, Neptune created the flying horse, Pegasus from a mixture of Medusa's blood and sea foam. Perseus flew on Pegasus along the Ethiopian coast, riding victoriously to rescue Princess Andromeda after killing the monster serpent Cetus threatening her.

King Cepheus and Queen Cassiopeia happily gave their blessings, and Andromeda sailed as bride with Perseus to his home island.

Upon returning Perseus learned his mother had rejected Polydectes' marriage proposal and the king then drove Danae and Dictys into hiding. Perseus tracked down the king and his minions in the middle of a great palace banquet. Standing at the entrance, wearing Athena's glorious armor, Perseus caught everyone's eye. He then withdrew the Medusa's head from its sack, turning the crowd into stone. Here's a reminder, Superhero Stars may be out of sight, but they best be not be forgotten!

After the islanders had gratefully accepted Perseus' recommendation and made Dictys their new king; Perseus, Andromeda, and Danae returned to Greece in hopes of being reconciled with Acrisius who had set the young mother and child adrift so many years before. However, they learned Acrisius had been sent into exile and no one knew where he was.

Hearing of a great athletic festival, Perseus journeyed to a northern realm to take part in the games. After throwing a discus, the projectile landed in a crowd and killed one of the spectators. It turned out the dead man was Perseus' exiled grandfather, who had been visiting the local king. With the death of Acrisius the Delphic prediction had finally come true.

Perseus and Andromeda went on to a happy life together, and the myths tell us they were the great-grandparents of the mighty hero Hercules – another superstar born for the great moving screen of sky theater!

VISUAL ASPECTS:

Perseus is in a lovely section of Milky Way when seen through binoculars.

An asterism known as the "Segment of Perseus" consists of a J-shaped chain of stars, including Mirfak, which extends towards the northeast. Another pattern, a reversed J, stretches southward from Mirfak and includes the star Epsilon Persei. This constellation contains several interesting features, including the double star cluster h and Chi Persei, located near the W of Cassiopeia; and the eclipsing binary star Algol.

All the bright individual stars seen in our sky belong to the local Orion arm of the Milky Way Galaxy, the spiral arm that also contains our Solar System. Approximately 5,000 light years from us, far beyond stars that outline the constellation Perseus, astronomers have identified an adjacent Milky Way arm located towards the Galaxy's outskirts.

In the opposite direction, halfway across the sky, in the constellation Sagittarius, another neighboring arm is observed. The Sagittarius spiral arm is positioned in the direction of the Galactic center and is seen as a glowing band, blended light of its distant stars and nebulae. Although about the same distance from us, the Milky Way region in Sagittarius appears much broader and brighter than its counterpart in Perseus. Toward Sagittarius we look in the direction of far more stars and nebulae located between us and the Galactic center than are observed at the Milky Way's fringes toward the direction of constellation Perseus. On evenings of early October, you can trace the tapered shape of the Milky Way's Band while gazing from Sagittarius low in the southwestern sky towards Perseus rising in the northeast.

During 1953 astronomer William W. Morgan (1906-1994) and his colleagues found concentrations of blue supergiant stars in Perseus, at distances of 4,000 to 6,000 light years, indicating the presence of that Galactic spiral arm. Blue supergiant stars provide a means of locating spiral arms in galaxies because those stars have relatively brief life spans precluding survival and migration far from their birthplaces in concentrated hydrogen and dust clouds of the Galactic spiral arms. Blue supergiants therefore are seen only in or very near those arms.

Analogous distributions of spectral type B and A stars within about 2,000 light years of the Sun provide a rough outline of the Galactic arm where we live. An example of one of these marker stars is Zeta Persei, a third magnitude type B supergiant, about 750 light years from us.

Zeta is the most prominent member of a star group called the Zeta Persei Association containing several dozen very luminous spectral type O and B stars, whose proper motions indicate mutual expansion from a common birth site. The group covers a region of space about 100 light years in diameter containing patches of nebulosity that may be remnants of the cloud out of which those stars formed. This stellar set is similar to that which comprises the Alpha Persei Association, seen concentrated in the lovely array around Mirfak, brightest star in Perseus.

Algol October 4

Northeast East-Northeast East

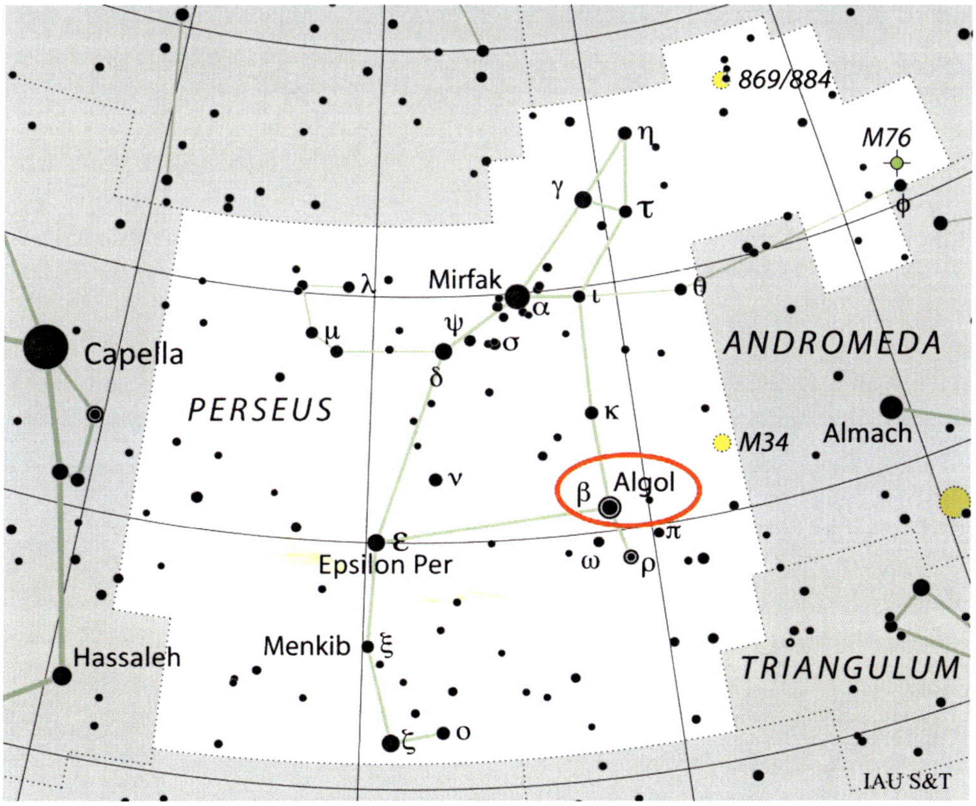

Perseus

October – March

AL-gall

2.1 mag. blue-white main sequence; 90 ly; prototype of the Algol type eclipsing binary variable stars

****

Imagine the Sun being the size of a baseball. At that scale, the two adjacent AB main component stars of Algol would each be about 10 inches wide. Those models would proportionally be at scaled distance from the baseball/Sun of about 28,000 miles, a separation equal to about 118% of the circumference of Earth.

If the Sun had the diameter of a Soccer ball, Algol AB models at this scale would be 77 centimeters wide. Scaled distance from the soccer ball/Sun to our Algol models would be about 134,500 kilometers, equivalent to three and a third times the great circle distance around Earth and 35% of the actual average distance from Earth to the Moon.

LOCATION:

This well-known variable star, normally of the second magnitude, periodically dipping to third magnitude, is located at a place in Perseus about one-third the angular distance from Mirfak to Hamal.

MEANING:

The name Algol is adapted from an Arabian description of this star as "Head of the Demon", which in turn was based on Ptolemy's characterization of this star as "Head of the Gorgon".

LORE:

Although no ancient or medieval references have been found specifically referring to Algol's light variation, early characterizations suggest its brightness variations had certainly been noted. Throughout the ancient Mediterranean world, Algol was usually associated with some form of dreadful monster.

Ancient Israelites are said to have referred to Algol as Lilith, Adam's legendary first wife, who apparently had frightful characteristics. In the classical sky legends of Greece and Rome, this star was seen as a representation of the "Head of Medusa", one of three loathsome sisters known as Gorgons, punished by the gods for their excessive vanity. Other traditional descriptions for Algol include "Demon Star", "Head of Medusa", "Eye of Medusa", and "Ghoul Star".

HISTORY:

Teenage English amateur astronomer John Goodricke (1764-1786) noted Algol's regular pattern of light variations and proposed an eclipsing binary mechanism to explain the changes in brightness. He's also credited with pioneering research of remarkable variable Delta Cephei – prototype of similar stars invaluable as standard candles for celestial distance estimations, and variable star Beta Lyrae. He overcame total deafness suffered since childhood and in 1786 was honored with the Copley Medal and membership in the Royal Society of London.

VISUAL ASPECTS:

Spectroscopic and interferometric analyses reveal the Algol system consists of three stars with an angular separation less than 0.016 arcsecond for component C from the considerably more snug eclipsing AB pair - whose separation is less than two milliarcseconds. Apparent magnitude of the Algol AB double system varies between 2.1 and 3.4 over a period of 2 days, 20 hours, and 49 minutes (2.867 days). Variability results from periodic eclipse interactions between the primary star and the slightly larger orbiting secondary.

Primary components of the eclipsing binary system, which are responsible for the Algol's apparent brightness variability, include Algol A (also known as Algol Aa1), a blue-white type B8 main sequence object with a luminosity about 94 times greater than that of the Sun. The secondary star, which produces the major eclipsing effect, is a spectral type K2 yellow-white subgiant approximately 3.4 times more luminous than the Sun, designated Algol B (aka Algol Aa2).

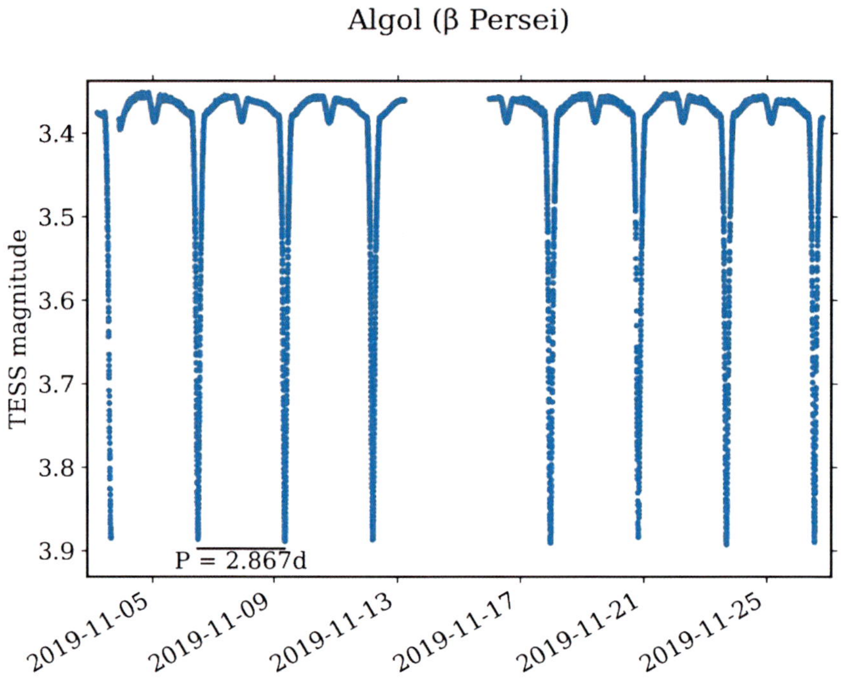

Algol light curve, data from NASA's Transiting Exoplanet Survey Satellite (TESS) Credit: Warrickball

PHYSICAL NATURE:

Algol Aa1 (Algol A) has a mass 3.4 times that of the Sun, surface temperature about 12,500 K, and its diameter is approximately 3.5 times that of the Sun, corresponding to a volume 43 times the Sun's size. The secondary, Algol Aa2 (Algol B), mass 0.8 solar, has a surface temperature of about 4,500 K. Although these stars are virtually the same size, the considerably greater temperature of the primary accounts for its much higher luminosity.

The C (Algol Ab) component is a spectral type F1 blue-white main sequence star with mass 1.8 solar and luminosity about 10 times the Sun's intrinsic brightness and orbits the mutually proximate AB star pair over a period of 680.2 days.

Nearness of Components Aa1 and Aa2 causes a gravitationally induced bridge of gas to extend out from secondary component B, forming a span known as a Roche lobe . Such structures have caused this and other stars similar to Algol to be known as semi-detached eclipsing binaries . The massive primary pulls bridging gas from its subgiant companion's atmosphere. Stars of the Algol Aa1/Aa2 pair are so close they became tidally locked, with one side of each always facing the other, thereby encouraging persistence of mass transferring action by means of the Roche lobe connection.

ALGOL - ADDITIONAL DATA: **Catalogue Designations: Beta Persei, β Per, Demon Star, El Ghoul, 26 Persei, HD 19356, HIP 14576, HR 936, SAO 38592********************

Parallax: 36.3 mas
Distance: 27.5 parsecs, 90 light years
****
Apparent magnitude: 2.1
Absolute magnitude: -0.1
Luminosity: 94 solar
Luminosity class: main sequence V
****
Apparent color: blue-white
Peak color: blue-white

Color index (B-V): -0.05
Spectral type: B8
Surface temperature: 12,500 K
****
Diameter: 3.5 solar
Mass: 3.6 solar
****
Coordinates: (J2000.0) 03h08m +40°57' / 47.04 +40.96
Transit at 9 p.m. December 24

Proper motion: +3 -2 mas/year
Radial velocity: +4 km/sec.
****

Epsilon Persei October 17

Northeast **East-Northeast** **East**

Perseus

October – March

EP-sigh-lon PURR-see-eye

2.9 mag. blue-white main sequence; 640 ly; variable star of the Beta Cephei type

****

If the Sun is visualized as being the size of a baseball, at that scale Epsilon Persei would be about 1.9 feet wide. The model would be about 199,400 miles from the baseball/Sun model, 83% of the Moon's actual distance from Earth.

With the Sun having the diameter of a Soccer ball, an Epsilon Persei model, at this scale, would be 1.7 meters wide. Scaled distance from the soccer ball/Sun to our metric Epsilon Persei model would be about 956,800 kilometers, two and a half times the actual distance between Earth and the Moon.

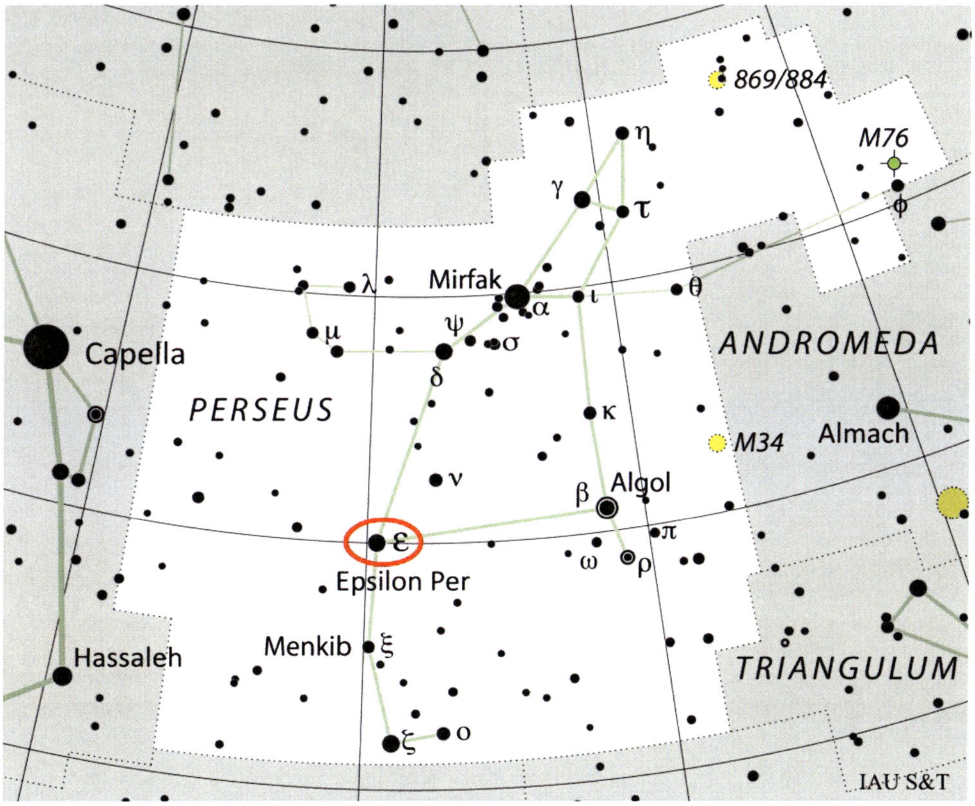

LOCATION:

Third magnitude Epsilon Persei is nine degrees east of Algol.

LORE:

Epsilon Persei was shown at the right knee of Perseus in the constellation art of Johannes Bayer's Uranometria, his watershed star catalogue from 1603.

HISTORY:

Epsilon Persei has an eighth magnitude stellar companion whose relative position was measured by Wilhelm von Struve in 1832. The pair has the same proper motion, but no evidence of orbital motion has been detected since Struve's time.

VISUAL ASPECTS:

Epsilon Persei is a blue-white member of the main sequence and is spectral type B0.5. It has a surface temperature of about 26,500 K and luminosity approximately 2,300 times that of the Sun.

The secondary star near Epsilon Persei is also a blue-white, main sequence bject, spectral type A2 and apparent magnitude 7.6. Its angular separation from the primary is nine arcseconds. Therefore, the companion is rather difficult to see in a small telescope due to considerable difference in brightness between it and the primary.

PHYSICAL NATURE:

Comprising a mass equal to 13.5 Suns, the B-V color index of Epsilon Per is -0.19; and it is 7.7 times the diameter of the Sun. Such girth represents a volume about 460 times larger than the Sun.

EPSILON PERSEI - ADDITIONAL DATA:

Catalogue Designations: Epsilon Persei, ε Per, 45 Persei, HD 24760, HIP 18532, HR 1220, SAO 56840, A1017

****

Parallax: 5.1 mas
Distance: 196 parsecs, 640 light years
****
Apparent magnitude: 2.9
Absolute magnitude: -3.6
Luminosity: 2,300 solar
Luminosity class: main sequence V
****

Apparent color blend: blue-white
Peak color: blue-white
Color index (B-V): -0.19
Spectral type: B0.5
Surface temperature: 26,500 K
****
Diameter: 7.7 solar
Mass: 13.5 solar

****
Coordinates: (J2000.0) 03h58m +40°01' / 45.97 +40.02
Transit at 9 p.m. January 5
Proper motion: +14 -24 mas/year
Radial velocity: -1 km/sec.
****

NEARBY FEATURES

About 10 degrees of arc southwest of Epsilon Persei is centered the Perseus Molecular Cloud System, at a distance of about 1,000 light years, it is one of the nearest great star-forming regions to the Sun. Distributed in this patch of interstellar dust and gas are over 150 Herbig-Haro (HH) objects - sites of new star formation.

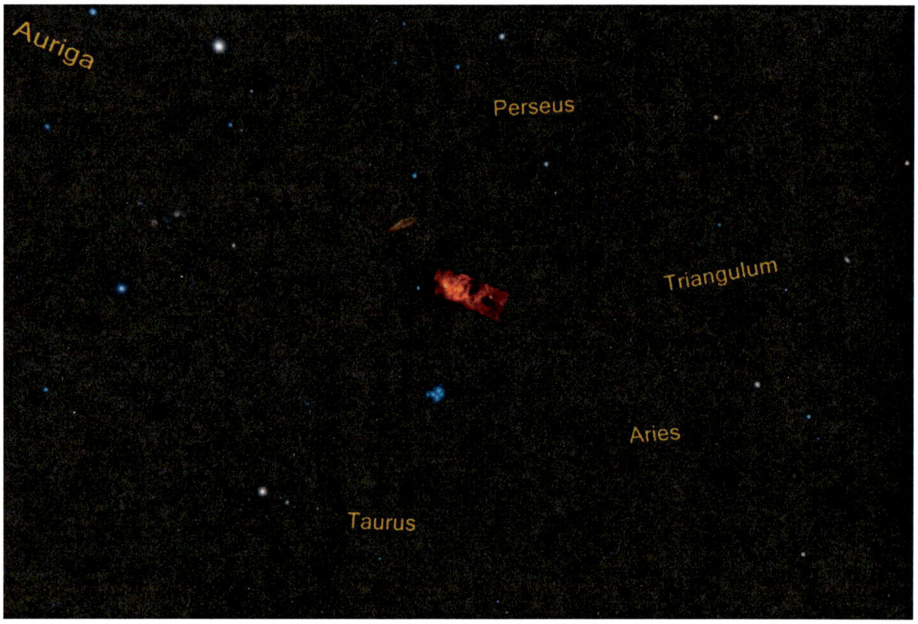

The Perseus Molecular Cloud's sky location relative to nearby constellations, is shown as a red inserted image from the Spitzer Space Telescope – Credit: NASA/JPL-Caltech

HH objects, also located in many other hot hydrogen HII star birthing sites (coincidence in nomenclature: HH = Herbig Haro object & HII = ionized hydrogen cloud at 20,000K+ temperature) around the sky such as the Great Nebula in Orion, are named after American astronomer George H. Herbig (1920-2013) of the University of California and Mexican astronomer Guillermo Haro (1913-1988), who first called attention to them. Investigations using infrared and radio telescopes suggest that most, if not all, protostars during their early episodes before joining the main sequence emit powerful bursts of gas from their surfaces which are most strongly able to penetrate relatively thin polar sections of surrounding rotation plane disks of dust and gas. Blasts of those bi-polar flows thrust way beyond the nascent stars out to as far as several light years.

The jets can pick up additional material from surrounding molecular dust clouds, driving that matter away from the fledgling star. Shock waves produced by those nebula interactions are generally invisible because their light is absorbed by molecular cloud dust. However, if a jet's shock wave reaches the surface of a cloud, which happens to be facing toward Earth, its light may escape in our direction and be observed as an Herbig-Haro object. HH12 (3h 28m 58s / 31° 20' 10") in Perseus is an example of such objects discovered by Herbig and Haro.

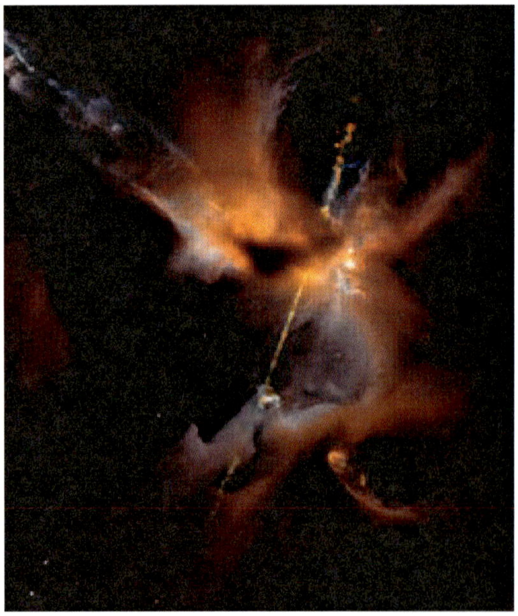

Needle-like bi-polar jets escaping north and south through a newly formed star's dark equatorial dust disk and on out into the surrounding bright HH24 object, located in the Orion B Molecular Cloud Complex, - Credit: HST / NASA/ESA

Fomalhaut October 21

South-Southeast South South-Southwest

Piscis Austrinus

October - November

FO-mal-oh

1.2 mag. blue-white main sequence; 25 ly

Anchor Point Star – A3 V

****

Imagine if the Sun was the size of a baseball, on the same scale Fomalhaut would be about five and a quarter inches wide. That model would have to be 7,800 miles from the baseball/Sun, approximately the distance from New York City to Antarctica.

On the other hand, if the Sun was the diameter of a Soccer ball, a Fomalhaut model, to a metric scale, would be 40 centimeters wide. Distance from our soccer ball/Sun to the Fomalhaut model would then be about 37,400 kilometers, 93% of a great circle flight around Earth.

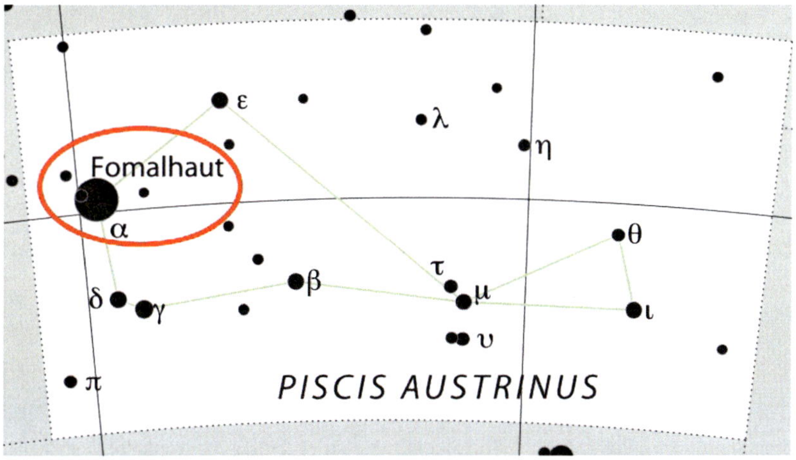

LOCATION:

Fomalhaut is a prominent first-magnitude star located 21 degrees southeast of Deneb Algedi, in a sparse stellar region.

MEANING:

In Ptolemy's Almagest Fomalhaut is described as "The Star in the Mouth" (of the southern fish). The name Fomalhaut is adapted from an Arabic translation of Ptolemy's second century A.D. portrayal. Another Arabian designation for Fomalhaut was "First Frog".

LORE:

Fomalhaut, along with Aldebaran, Regulus, and Antares were considered "Royal Stars" that served as "Guardians of Heaven" in ancient Persian sky lore. This star has sometimes been portrayed associated with water around the feet of Aquarius, and Ptolemy also called it, "That at the Beginning of the Water".

Due to a lack of bright stars in its neighborhood Fomalhaut is sometimes known as the "Solitary One". As one faces south and looks at the autumn evening sky, it's easy to come face to face with Fomalhaut. Its brightness and isolated position give it a distinguished identity as a symbol of the autumn season. With each passing year when you see Fomalhaut during weeks when trees are richly adorned in their fall colors, we believe the title "Autumn Star" will increasingly seem appropriate for this lovely evening feature of the cool southern sky.

VISUAL ASPECTS:

Blue-white Fomalhaut is a main sequence object with an apparent magnitude of 1.2. Its distance from Earth is 25 light years – one of the closest bright stars to the Solar System. Fomalhaut's surface temperature is estimated at 8,600 K, and it is about 17 times more luminous than the Sun.

PHYSICAL NATURE:

The A3V spectrum of Anchor Point star Fomalhaut is characterized by strong absorption lines of hydrogen and lines of ionized metals such as iron, titanium, and magnesium. Main sequence stars of Fomalhaut's luminosity are estimated to have masses about 1.9 times that of the Sun. Its diameter, 1.8 solar, indicates a volume 5.8 times the size of the Sun.

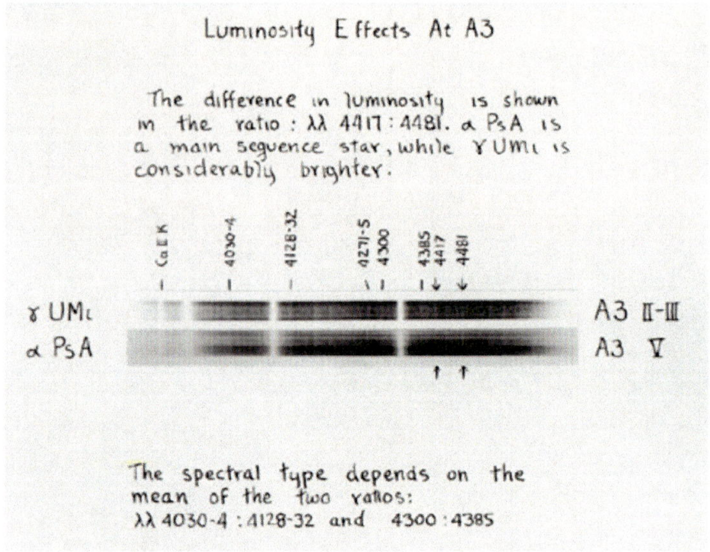

Chart showing subtle spectrum differences between "middle age" stars such as Fomalhaut and giant "senior citizen" evolved stars of similar surface temperature, such as Gamma Ursae Minoris. Credit: MKK Atlas

FOMALHAUT - ADDITIONAL DATA:

Catalogue Designations: Alpha Piscis Austrini, α PsA, 24 Piscis Austrini, HD 216956, HIP 113368, HR 8728, SAO 191524

****

Parallax: 129.8 mas

Distance: 7.7 parsecs, 25 light years

****

Apparent magnitude: 1.2

Absolute magnitude: + 1.7

Luminosity: 17 solar

Luminosity class: main sequence V

****

Apparent color blend: blue-white

Peak color: white

Color index (B-V): 0.09

Spectral type: A3

Surface temperature: 8,600 K

****

Diameter: 1.8 solar

Mass: 1.9 solar

****

Coordinates: (J2000.0) 22h58m -29°37' / 344.41 -29.62

Transit at 10 p.m. October 21

Proper motion: +329 -165 mas/year

Radial velocity: +7 km/sec.

****

NEARBY FEATURES:

Dagon (Fomalhaut b) - for a while was considered to be an exoplanet, but now is believed to have been a dust cloud orbiting Fomalhaut.

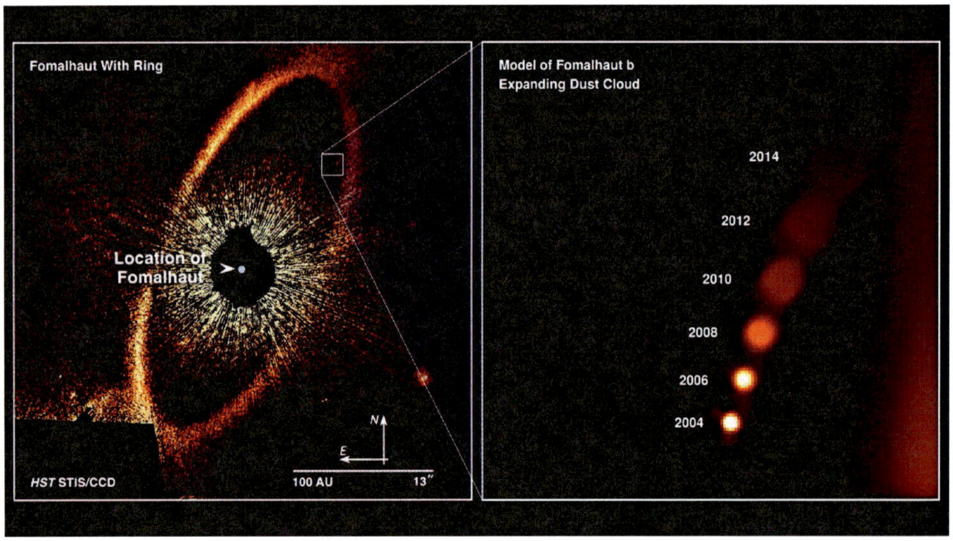

In 2005 it was announced that Fomalhaut is surrounded by a substantial ring of dust. Paul Kalas and James Graham, both at U. California, Berkeley, and Mark Clampin, at Goddard Space Flight Center, reported in *Nature* that Fomalhaut is encompassed by a disk of microscopic grains whose plane of revolution is tilted at an angle of 24° to our line of sight. The disk's inner edge is sharply defined at a distance of 133 astronomical units (A.U.) from that star and it is 25 A.U. wide. For comparisons, Earth's average distance from the Sun, about 93 million miles, defines one A.U. Jupiter's solar distance is 5.2 A.U. and Uranus has an average orbital radius of 19 A.U.

Rather than centered on Fomalhaut itself, the ring's geometric middle is offset by a substantial 15 A.U. from the star. This asymmetry prompted Alice Quillen of the University of Rochester to suggest one or more planets were in orbit around Fomalhaut. Kalas also led a team at Berkeley that spotted the supposed planet on images produced by the Hubble Space Telescope's Advanced Camera for Surveys. They announced in 2008 that a planet designated Fomalhaut b with a mass no more than three times that of Jupiter orbited in the vicinity of the dust disk.

NASA's Exoplanet Archive lists the planet Dagon (Fomalhaut b) with a mass about 2.6±0.9 times Jupiter's, semi-major axis distance from Fomalhaut about 115 A.U., and indicates an orbital period of about 875 years.

However, analysis of Hubble Space Telescope data from 2013 and 2014 showed Fomalhaut b was becoming less bright and was expanding, suggesting it was a debris cloud resulting from the violent collision of two asteroids which had been orbiting Fomalhaut. Erstwhile planet Fomalhaut b was canceled from NASA's Exoplanet Archive in 2020, and by 2023 had disintegrated to the degree that it could not be seen in James Webb Telescope images. R.I.P. Fomalhaut b!

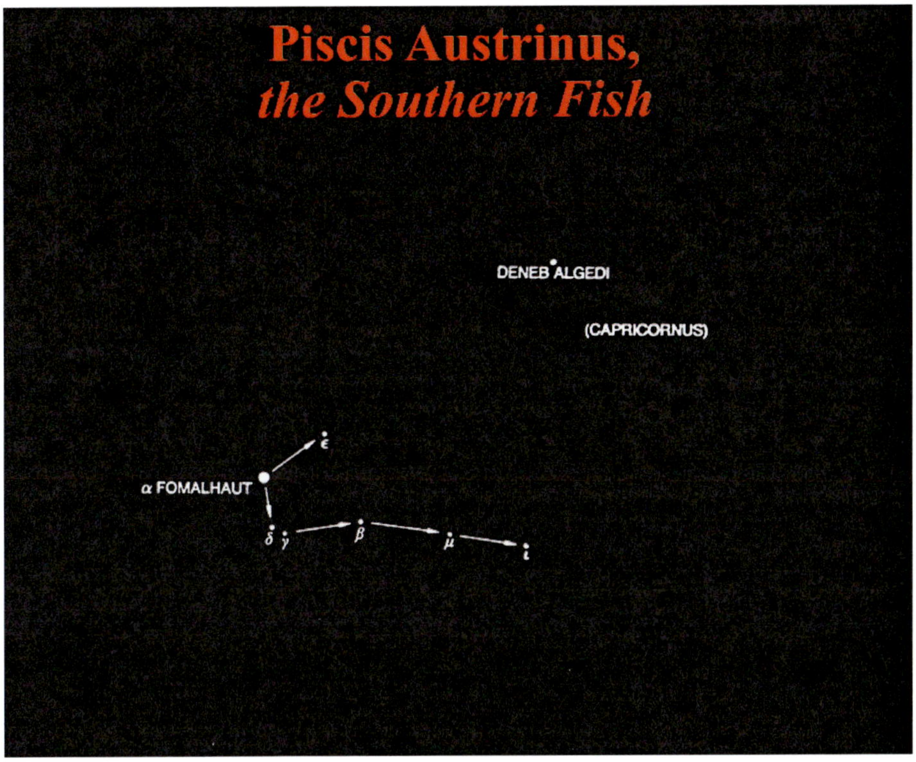

PICE-iss (PIES-iss) aw-STRY-nus
PsA – Piscis Austrini,
PICE-iss (PIS-iss) aw-STRY-nye
>>>>>>>>>>>>>>>>>>>>>>
Brightest Star
October 21 Fomalhaut 1.1
blue-white main sequence 25 ly
>>>>>>>>>>>>>>>>>>>>>>
EVENING SEASON:
October - November
SKY TRACK:
Stars of Piscis Austrinus rise in the southeast, transit the meridian low in the southern sky, and set towards the southwest.
LORE:
Piscis Austrinus is one of several constellations in this part of the sky associated with water. Traditional pictures of the constellation figure often show the Southern Fish with its mouth open, drinking from a flow pouring from the "Water Jar" of Aquarius.

VISUAL ASPECTS:
Piscis Austrinus is southeast of Capricornus and directly south of Aquarius.
This faint constellation is dominated by its single bright star Fomalhaut; which has apparent magnitude 1.1. The constellation's next brightest star is 16 times fainter with a magnitude of only 4.2. For this reason, the elongated figure of stars that traditionally outlines the Southern Fish is best seen with binoculars.

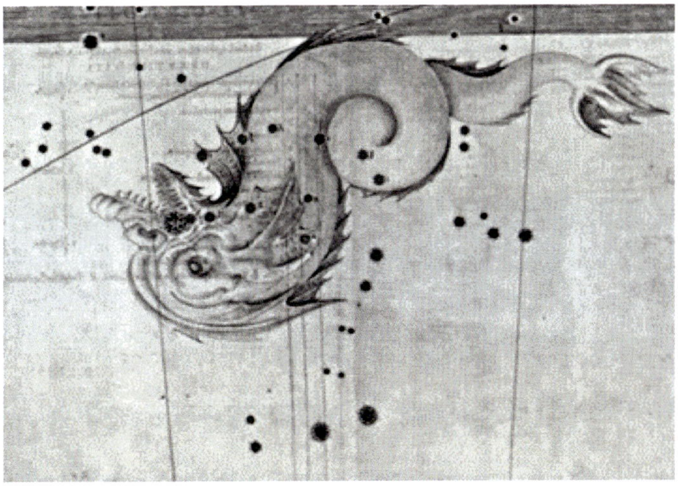

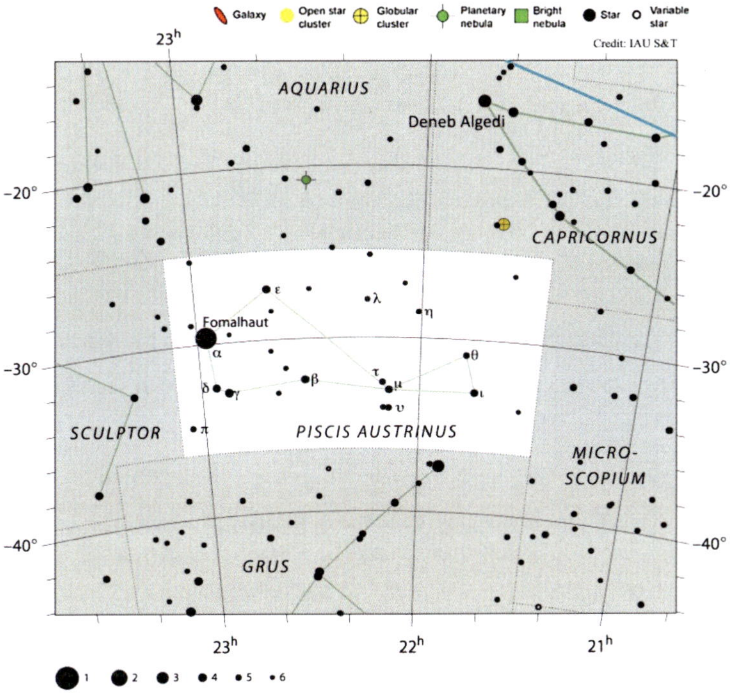

Diphda October 22

Southeast South-Southeast South

Cetus

November – January

DIF-dah

2.0 mag. yellow giant; 96 ly

****

Imagine the Sun being the size of a baseball. On that scale, Diphda would be 4.1 feet wide– about a quarter the width of a pitcher's mound. That model would have to be at a distance from the baseball/Sun of about 29,900 miles, equal to 120% of earth's circumference.

If the Sun had the diameter of a Soccer ball, a metric Diphda model at this scale, would be 3.7 meters wide – one-fifth the diameter of a playing field's center circle. Scaled distance from the soccer ball/Sun to our Diphda model would be about 143,500 kilometers, 37% of the actual average distance from Earth to the Moon.

LOCATION:

Twenty-eight degrees northeast of Fomalhaut is Diphda, at second-magnitude it's the brightest star in constellation Cetus.

MEANING:

The name Diphda was applied during the early 19th century and derives from the Arabian designation "Second Frog". Fomalhaut in Piscis Austrinus represented the "First Frog".

LORE:

In sky descriptions of ancient Greece and Rome, Diphda marked the southern branch of the "Whale's Tail". An alternate name Deneb Kaitos is from an Arabian translation of that classical description. Astronomers in medieval China knew Diphda as "Superintendent of the Earthworks".

VISUAL ASPECTS:

This is a yellow giant with apparent magnitude 2.0. At a distance of 96 light years from the Solar System, its luminosity is 117 times greater than the Sun's.

PHYSICAL NATURE:

With spectral type K0, Diphda has a surface temperature estimated at approximately 4,800 K. Its luminosity class is class III giant with a diameter 17 times the solar width. That diameter indicates a volume 4,900 times larger than the Sun.

Absorption lines of neutral metals dominate its spectrum. The right ascension and declination vectors of this star's proper motion are +233 milliarcseconds/year toward the east and 32 mas northward per year. The positive direction of right ascension indicates eastward proper motion. The positive direction of declination indicates a northward proper motion. Diphda's distance from the Sun increases by 13 kilometers every second.

DIPHDA - ADDITIONAL DATA:

Catalogue Designations: Beta Ceti, β Cet, Deneb Kaitos, 16 Cet, HD 4128, HIP 3419, HR 188, SAO 147420

****

Parallax: 33.9 mas

Distance: 29.5 parsecs, 96 light years

****

Apparent magnitude: 2.0

Absolute magnitude: -1.3

Luminosity: 283 solar

Luminosity class: giant III

****

Apparent color blend: yellow

Peak color: orange

Color index (B-V): 1.01

Spectral type: K0

Surface temperature: 4,800 K

****

Diameter: 17 solar

Mass: 2.8 solar

****

Coordinates: (J2000.0) 00h44m -17°59' / 10.90 -17.99

Transit at 9 p.m. November 17

Proper motion: +233 +32 mas/year

Radial velocity: +13 km/sec.

****

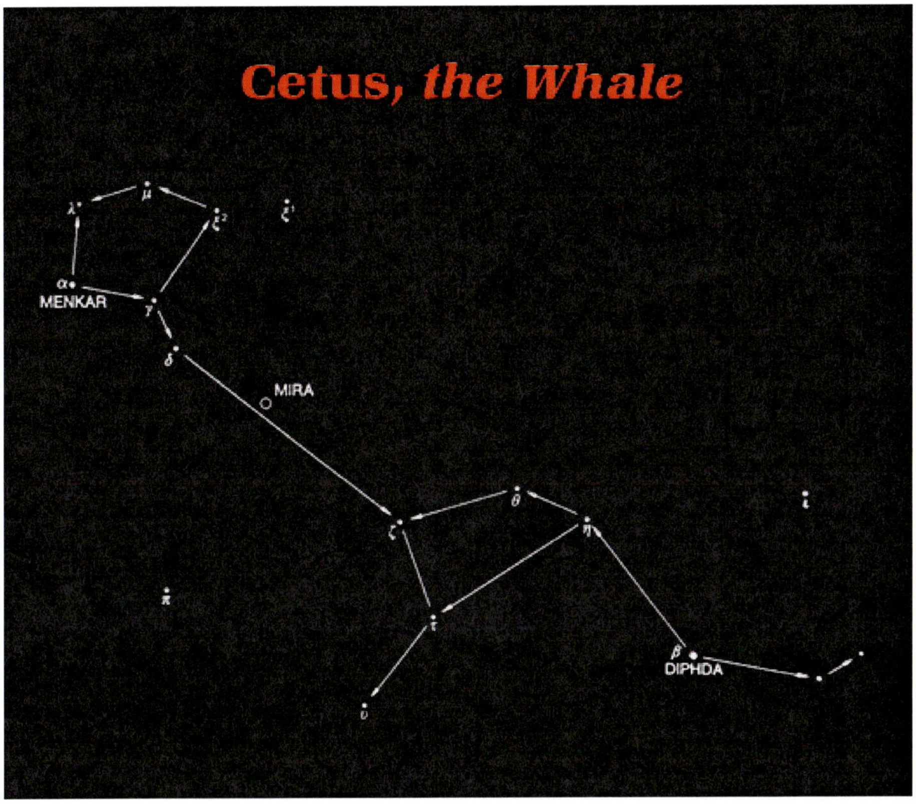

SEE-tus

Cet – Ceti, *SEE-tie*

\>>>>>>>>>>>>>>>>>>>

Brightest Stars

October 22 Diphda 2.0 yellow giant 96 ly

October 27 Menkar 2.5 yellow-orange giant 250 ly

\>>>>>>>>>>>>>>>>>>>

EVENING SEASON:

November - January

SKY TRACK:

Cetus rises in the southeast, crosses the meridian less than halfway from horizon to zenith, and sets along the southwestern horizon.

LORE:

Since at least the time of Hellenic writer Aratus in the third century B.C., Cetus has been associated with the "Sea Serpent" sent to the coast of Ethiopia by gods of Olympus as punishment to Cassiopeia. That creature was eventually slain by Perseus when he rescued Andromeda.

Although the title "Whale" has been applied to this constellation since before Roman times, traditional representations of the creature's figure look more like imagined portrayals of the Loch Ness Monster than of any known cetacean.

A circlet of six stars extending to the northwest from Menkar defines the creature's head, connected to its body by a very long neck. The smallish body of Cetus is figured from stars including Diphda and several of its eastern neighbors.

VISUAL ASPECTS:

With its relatively faint stars, this constellation covers an extended portion of the sky; consequentially its outline is rather difficult to trace unless observing conditions are fine. Cetus is south of the celestial equator, beneath Andromeda and Pisces as they cross the meridian. Its position is close to the South Galactic Pole. As we look toward this constellation, we're viewing sparse hinterlands, stellar "Ultima Thule" perpendicular to the Galactic plane; 90 degrees of arc from the Band of the Milky Way. We therefore gaze at Cetus away from the direction of galactic nebulae and areas of active star formation such we seen in Scorpius and Orion. In those latter regions groups of massive highly luminous young stars are visible, and new generations of brilliant stars are presently being formed. On the other hand, at the sky region of Cetus the South Celestial Poles offers a viewing portal out toward intergalactic space and rich fields of the myriad galaxies beyond our own.

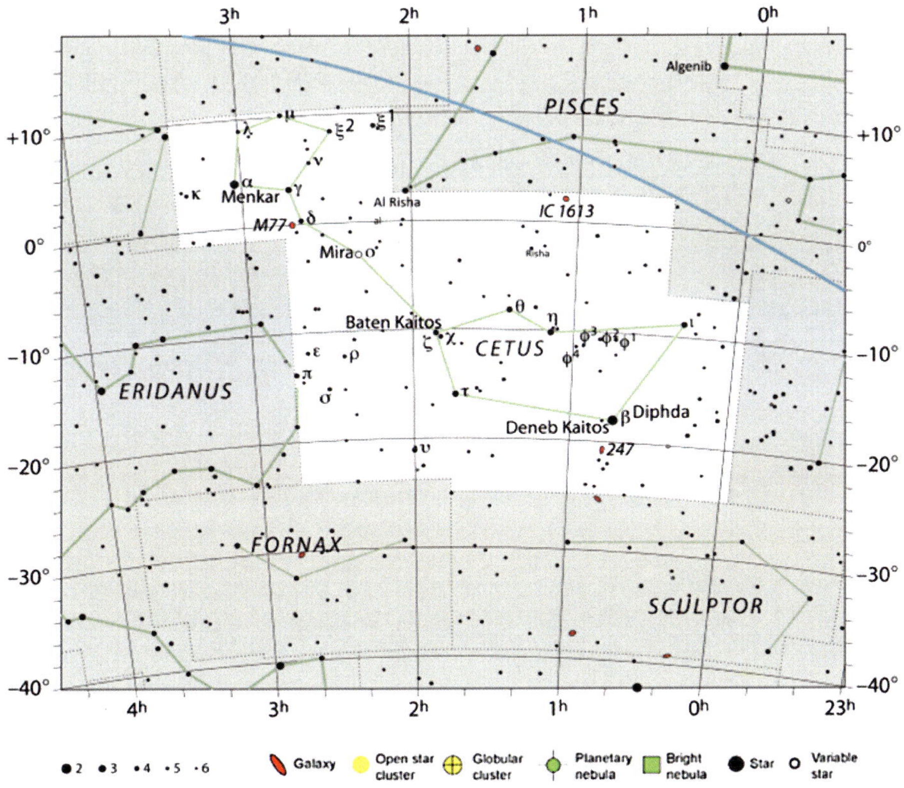

Credit: IAU S&T

Alcyone October 24

East-Northeast East East-Southeast

Taurus

November – March

al-SIGH-own

2.9 mag. blue-white giant; 446 ly

****

Imagine the Sun being the size of a baseball. At that scale a model of Alcyone would be almost two feet wide, and 124,600 miles from our baseball/Sun. Such separation is equal to 52% of the Moon's actual distance from Earth.

On the other hand, using a metric comparison, if the Sun had the diameter of a Soccer ball, an Alcyone model to scale would be 1.8 meters wide. Proportional distance from our soccer ball/Sun to an Alcyone model would be about 598,000 kilometers, 1.6 times the average distance between Earth and Moon.

LOCATION:

Third magnitude Alcyone is the brightest star of the Pleiades, that lovely archetypical open star cluster in Taurus. Alcyone is 16 degrees south of Epsilon Persei, also a third magnitude object..

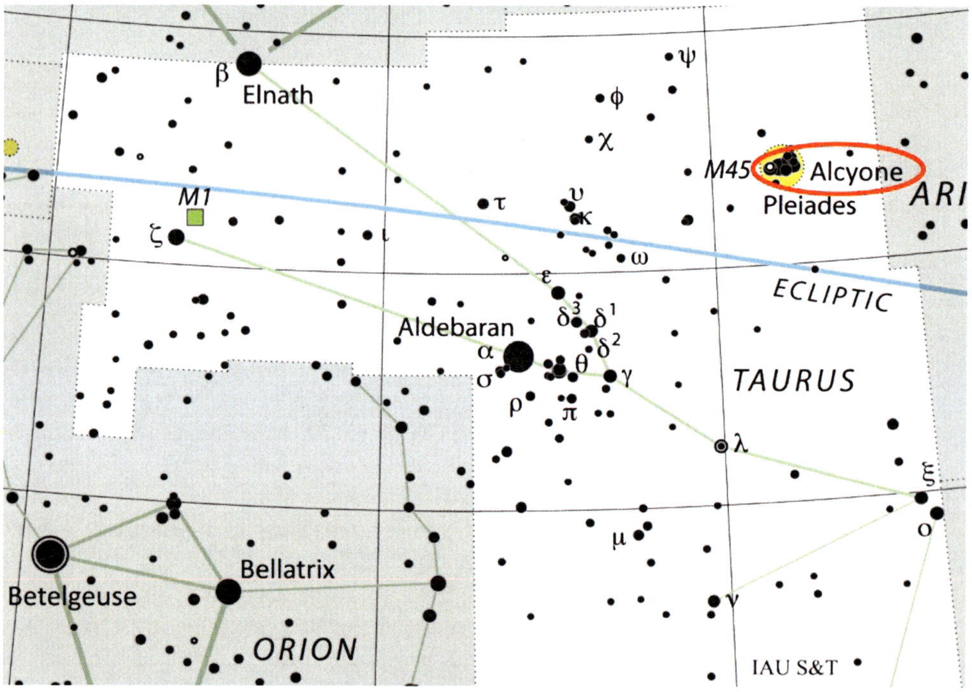

MEANING:

Alcyone is named after one of the "Seven Sisters of the Pleiades", daughters of Atlas, and half-sisters of the Hyades in Greek mythology.

LORE:

Stars of the Pleiades form what likely is the best-known of all stellar clusters. During the age of classical Greece, rising of the Pleiades into the predawn eastern sky coincided with arrival of fair weather and start of navigation season on the Mediterranean. As a result, the Pleiades came to be known as the "Sailors' Stars", a title which remained popular in England and Germany up to recent times.

Many cultures around the world had special names for stars of the Pleiades. Anglo-Saxons and medieval Germans called it the "Seven Stars". In Finland and Lithuania, the Pleiades were known as a "Net", and in Russia, they were known as a "Sitting Hen". Swedish sky watchers used the description "Fur in Frost", a poetic comment on the sparkling delicacy of these stars. In Denmark, the name "Evening Hen" was used, and in Greenland Eskimos called the Pleiades a "Pack of Dogs". A similar description, "Close Pack", was used in Wales.

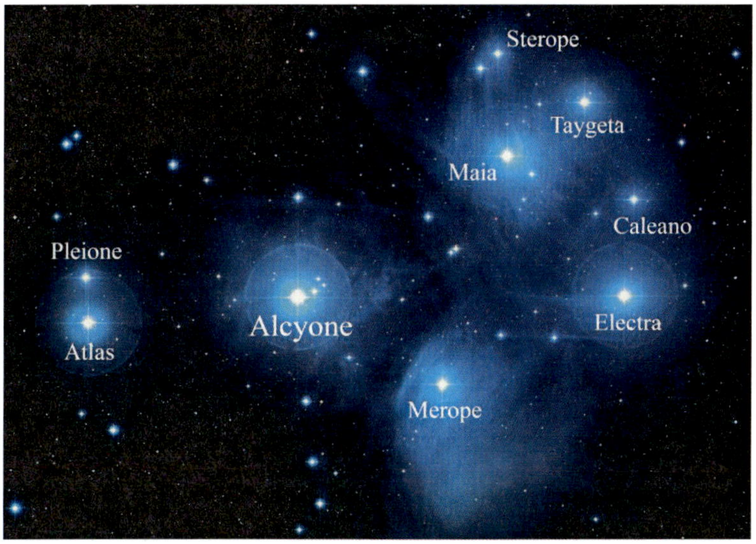

The Pleiades, M45 – Open Star Cluster Credit: NASA, ESA, AURA/Caltech, Palomar Observatory

Although the Pleiades are often called the "Seven Sisters", only six of its stars are usually seen distinctly by average visioned observers on clear dark nights. On the other hand, sharp-eyed sky watchers with excellent observing conditions report seeing more than seven stars.

Proper motion studies indicate the Pleiades cluster contains several hundred members. It was listed as number 45 on Messier's roster of celestial objects and often referred to as M45. It's one of the nearest star clusters to Earth, with distance to its center being about 446 light years.

Knowledge of that span enabled calculations of absolute magnitudes for stars in the cluster. That in turn allows correlations to be made between those values and colors of the stars plotted on an H-R diagram. With such a chart in hand, main sequence profiles obtained for more distant stellar clusters may be compared in a technique called "main sequence curve fitting", which provides estimates of distances to those more remote clusters. That method, historically often used in studies of the closer Hyades star cluster also in Taurus, provided an important tool for astronomers, especially before far reaching highly accurate distance measurements from the European Space Agency's Gaia mission became available beginning in 2016 and 2018.

Position of the upper left-hand terminus of a star cluster's main sequence suggests to astronomers the approximate time since stars in the cluster formed out of Milky Way dust and gas. Current left end of the Pleiades main sequence plot indicates this cluster's age is about 120 million years.

A nebulous haze enveloping stars of the Pleiades visible in small telescopes is composed of dust particles reflecting light from the cluster's stars back toward us. That nebulosity is most evident around the Pleiad Merope. Dark absorption lines in the nebula's spectrum provided initial evidence that some nebulae shine by reflected starlight, and not by any intrinsic heat induced glow.

Previously it was believed nebulosity surrounding the Pleiades was debris left over after formation of the cluster's stars. However, cluster stars just happen to be passing through a dust cloud which earlier formed independently.

The Pleiades are a spectacular sight in binoculars, and its brightest stars may be seen to resemble a tiny dipper shaped asterism.

A lovely reference to the Pleiades was made by Alfred, Lord Tennyson (1809-1892), in his poem *Locksley Hall*:

"Many a night I saw the Pleiads, rising thro' the mellow shade,

Glitter like a swarm of fireflies tangled in a silver braid".

HISTORY: In 1610, Galileo published the first telescopically based drawing of Alcyone and other Pleiades stars in his Sidereus Nuncius (Starry Messenger). Charles Messier's 1771 catalogue of comet resembling non-comets includes the Pleiades as number 45. However, through unaided 20-20 eyes the Pleiades appear as a dipper shaped formation of individual stars, not the hazy spot seen when observing comets or diffuse nebulae

VISUAL ASPECTS:

Alcyone, brightest of the Pleiades, has an apparent magnitude of 2.9. It's a blue-white giant 446 light years from the Solar System. The star has surface temperature 12,800 K and luminosity about 890 times greater than our Sun.

PHYSICAL NATURE: The stellar system we perceive with unaided eyes as Alcyone consists of at least four members, brightest of which, designated Alcyone A, is a B7 III giant with mass six times the Sun's and a diameter eight times greater than the Sun. This primary component of Alcyone has a volume about 512 times that of the Sun.

Component B – designated 24 Tauri is a 6.28 magnitude white spectral type A0 main sequence star at an angular distance of 117 arcseconds from the primary.

A Delta Scuti type variable star with a range of magnitude 8.25 to 8.30 and period 1.13 hours is known as component C in Alcyone's multiple star system.

Two type F3 main sequence stars with a combined magnitude of 9.15 are together designated as component D. Four additional fainter system members are indicated in the Washington Double Star Catalogue published by the United States Naval Observatory.

ALCYONE - ADDITIONAL DATA:

Catalogue Designations: Eta Tauri, η Tau, 25 Tauri, HR 1165, HD 23630, HIP 17702, SAO 76199

****

Parallax: 7.3 mas
Distance: 137 parsecs, 446 light years
****
Apparent magnitude: 2.9
Absolute magnitude: - 2.6
Luminosity: 890 solar
Luminosity class: giant III
****

Apparent color blend: blue-white
Peak color: blue-white
Color index (B-V): -0.09
Spectral type: B7
Surface temperature: 12,800 K
****
Diameter: 8 solar
Mass: 6 solar

****
Coordinates: (J2000.0) 03h48m +24°09' / 56.87 +24.11
Transit at 9 p.m. January 2
Proper motion: +19 -44 mas/year
Radial velocity: +5 km/sec.
****

NEARBY FEATURE: Teide 1: the first brown dwarf identified (3h47m +24°22')

Shiv S. Kumar (1939-) working at NASA and the National Research Council in 1962 theorized about relatively low mass sub-stellar objects – spheres of hydrogen and helium which lack sufficient gravitational crushing force to have initiated and sustain proton-proton thermonuclear reactions in their cores. Without that primary mode of stellar "fire", insufficient amounts of heat and visible light can be produced for them to be considered "real" stars.

Two decades before any of such objects were actually discovered, they were dubbed "brown dwarfs" at the suggestion of astronomer Jill Tarter (1944-). Teide 1 was the first entity conclusively identified as a brown dwarf. In 1995 it was found in the heart of the Pleiades open cluster by Rafael Rebolo Lopez (1961-) and his team at the Instituto de Astropisica de Canarias, Tenerife, Spain. The object was named after Teide Observatory, where it was spotted during a photometric survey of the Pleiades.

When looking at this famous star cluster with binoculars or telescope, in your mind's eye consider the site of brown dwarf Teide, near Maia, the "top right" star of the bowl as you visualize stars of the Seven Sisters forming a tiny Dipper.

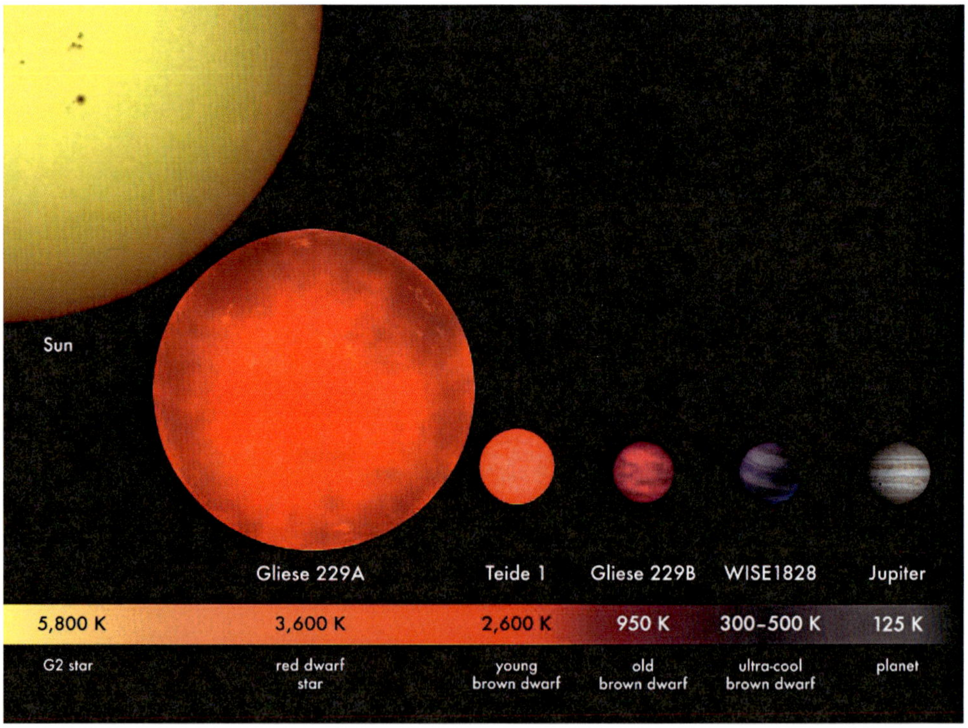

Image Credit: MPIA/V. Joergens

Because its proper motion and radial velocity match-up with cluster members including Alcyone and Maia, Teide is identified as a member of the Pleiades. Thereby knowing its distance and apparent faintness reveals Teide has surprisingly low luminosity (for a true star), inferring it's too small to trigger and maintain proton-proton fusion reactions. That limitation and spectral profile fit the definition of a brown dwarf.

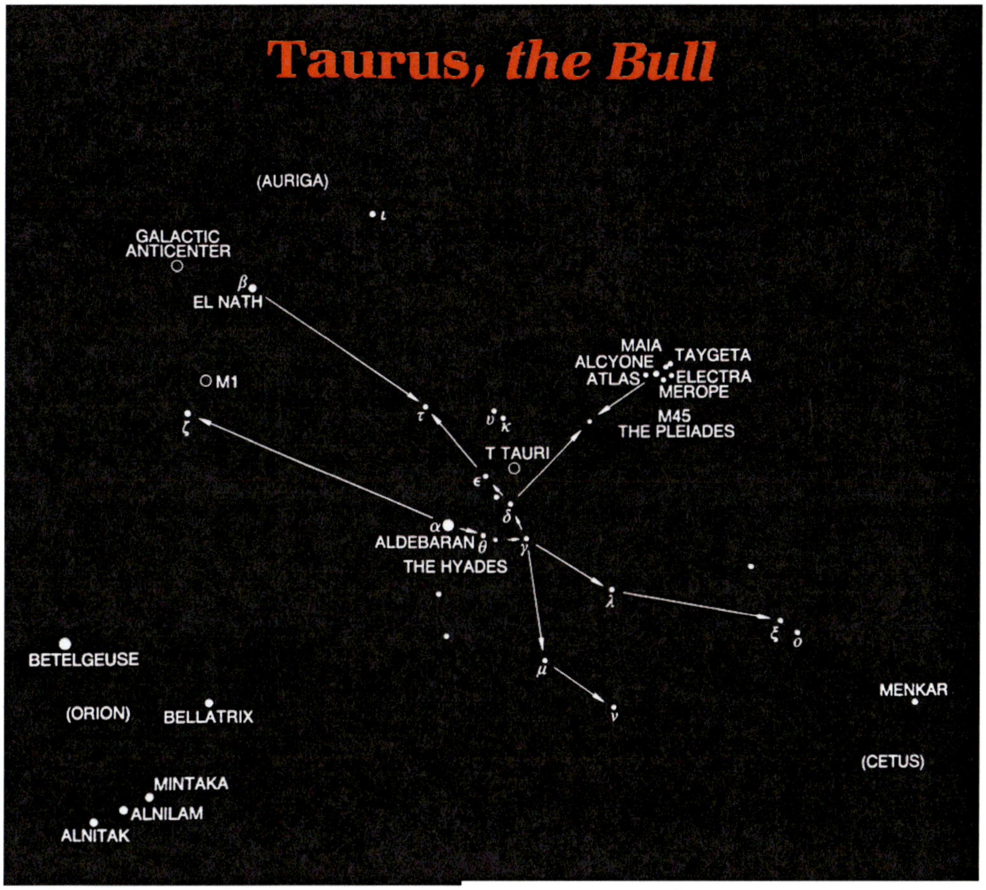

TOR-us

Tau – Tauri, TOR-eye

>>>>>>>>>>>>>>>>>>>>>

Brightest Stars

October 24 Alcyone 2.9 blue-white giant 400 ly

November 10 Aldebaran 0.9 yellow giant 67 ly

November 15 El Nath 1.7 blue-white giant 140 ly

>>>>>>>>>>>>>>>>>>>>>

EVENING SEASON:

November - March

SKY TRACK:

Taurus rises in the northeast, crosses the meridian high in the southern sky, and sets along the northwestern horizon.

LORE:

Taurus was one of the earliest groups of stars to be considered a constellation. The Bull, a symbol of the vitality of spring, served to mark the Sun's position in the sky at the start of spring during a period from about 4000 B.C. to 1700 B.C. Since then, precessional motion of Earth carried the Vernal Equinox westward through the constellation Aries to its present location in Pisces.

In Hellenic mythology, Taurus was identified with a Bull whose form Zeus assumed as he tried to woo the maiden Europa. Traditional pictures of Taurus often portray him swimming away through the sea with Europa on his back.

VISUAL ASPECTS:

Taurus' face is represented by the V-shaped pattern of stars formed by the Hyades open star cluster. First-magnitude Aldebaran, although not a cluster member, marks the Bull's fiery eye. The Hyades cluster is centered about 135 light years from Earth, according to estimates based on data in the Gaia 2 release.

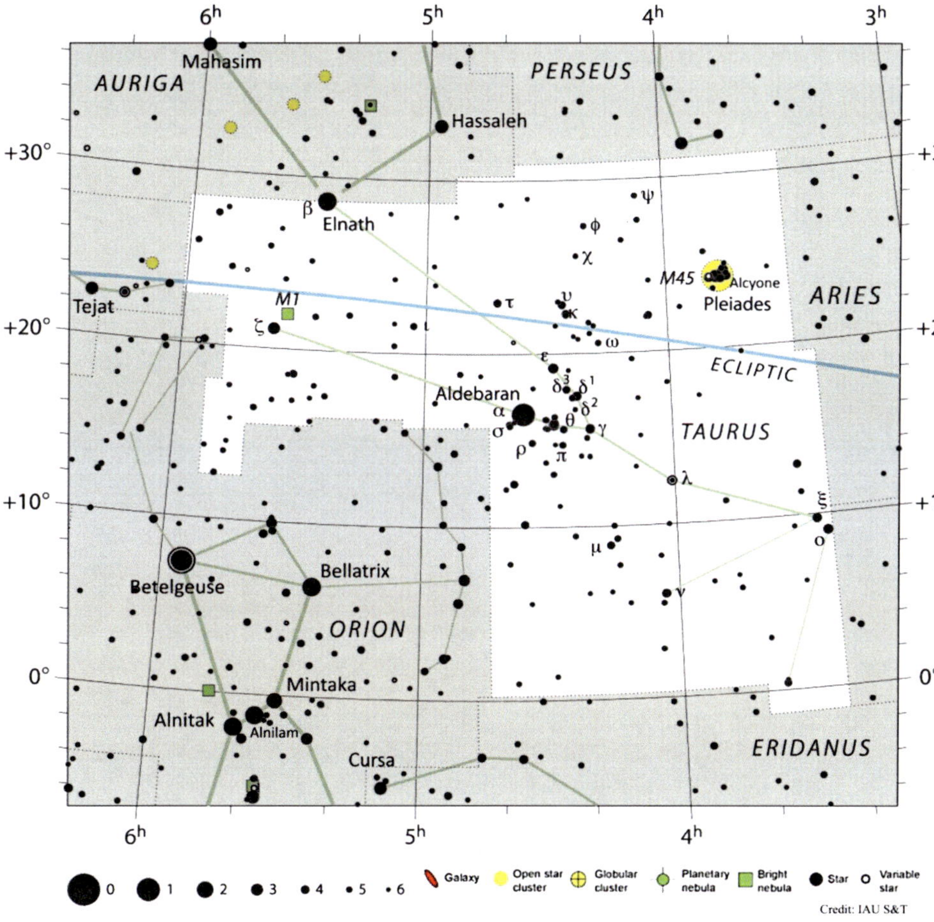

Before that European Space Agency mission, the Hyades were of prime importance in the hierarchy of techniques used to determine distances in space. Along with the Pleiades, yhe Ursa Major moving cluster and the Scorpius-Centaurus cluster, the

Hyades are near enough to Earth so that direct trigonometric techniques, such as the "moving cluster" method, could be used to measure its distance. Range estimates to more remote open star clusters were then inferred by comparisons of their H-R diagrams to those of the Hyades and other open clusters which are relatively near us.

The Pleiades, aka Seven Sisters, open star cluster is centered about 10 degrees northwest of the Hyades. Stars of the Pleiades often resemble a tiny cloud when seen with the naked eye. They can be resolved into stars with the unaided eye if one has excellent vision, and are a wonderful sight in binoculars and low-power telescopes. Alcyone is the brightest member of this remarkable, iconic stellar group.

NEARBY FEATURE:

***T Tauri - prototype for a class of pre-main sequence stars* (4h 22m +19° 32') Advancements in radio, infrared, and ultraviolet observing techniques enabled astronomers to begin understanding a class of objects known as T Tauri stars. The irregular variable star designated T Tauri is has tenth-magnitude and is located about five degrees northwest of Aldebaran.**

This star varies unpredictably in brightness between the ninth and thirteenth magnitude. The associated nebula has brightened considerably since the 1920s, and it appears to be directly influenced by T Tauri. During the 1940s, other stars similar to T Tauri were found by Alfred H. Joy (1882-1973) of the Mount Wilson Observatory in neighboring regions of Taurus and Auriga, as well as within nebulae Messier 16 in Serpens, M42 in Orion, and NGC 2264 in the constellation Monoceros.

T Tauri type stars all display erratic light variations of up to three magnitudes, intense spectral emission lines, and associations with dust laden molecular clouds. T Tauri stars also show evidence that they are ejecting considerable quantities of mass into ambient space at high speeds.

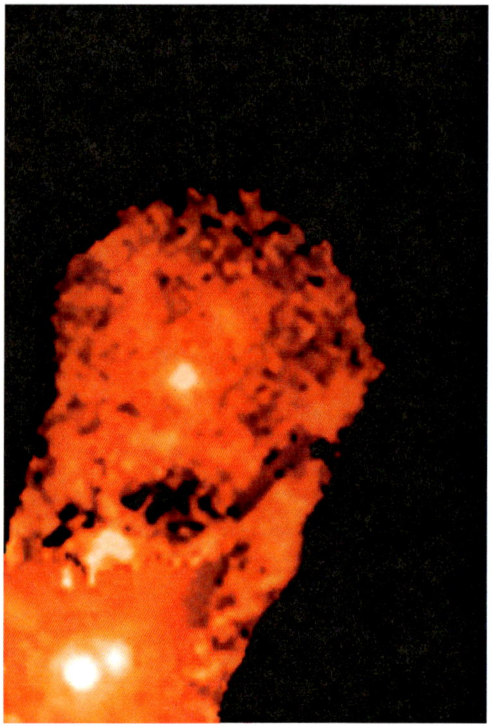

Image: upward to right surface eruption from binary T Tauri type star XZ Tauri – Credit: NASA/JPL/Jeff Hester (Arizona State University), Chris Burrows (ESA/STScI)

When luminosities and temperatures observed for T Tauri stars are plotted on a Hertzsprung-Russell diagram, it appears they are all very young, still in the process of gravitational contraction. They have not yet "reached" the main sequence zone of an H-R diagram, the stage when thermonuclear fusion of hydrogen to helium and energy takes place in stellar cores.

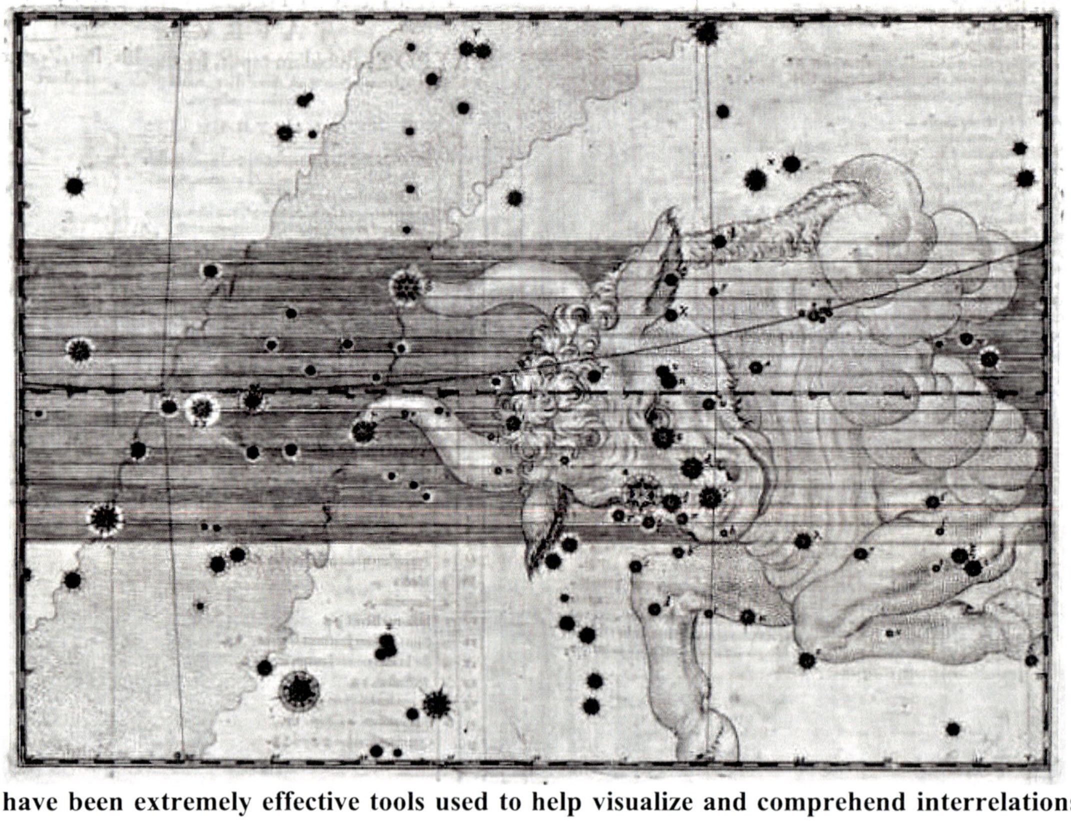

An Hertzsprung-Russell diagram (also known as a color-luminosity diagram) is a graph representing a star's color (as surrogate for surface temperature) along the x-axis on its lower edge and the star's brightness (either intrinsic or how bright is appears when viewed from Earth) arrayed on the Y - vertical edge. Since its introduction early in the 20th century based on research of Danish astronomer Ejnar Hertzsprung and American Henry Norris Russell, many variations of H-R diagrams have been extremely effective tools used to help visualize and comprehend interrelationships between various types of stars and their evolution.

Menkar October 27

East **East-Southeast** **Southeast**

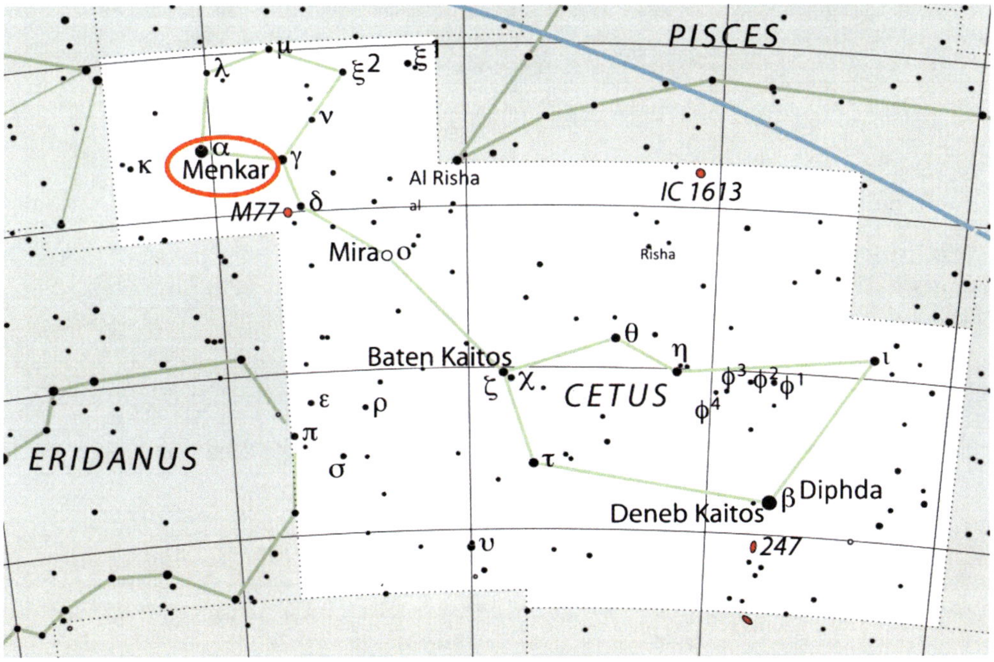

Cetus

November – January

MEN-kar

2.5 mag. yellow-orange giant; 250 ly

****

Visualizing the Sun as the size of a baseball, Menkar in comparison would be about 21.5 feet wide – 1.2 times wider than a pitcher's mound. This model would have to be about 77,900 miles from the baseball/Sun – 33% of the Moon's actual distance from Earth to be in scale.

If the Sun had the diameter of a Soccer ball, a Menkar model, at this metric scale, would be 19.6 meters wide –107% the size of a soccer field's center circle. Scaled distance from the soccer ball/Sun to this Menkar model would be about 373,800 kilometers, 97% of the average distance from Earth to the Moon.

LOCATION:

Third magnitude Menkar in the constellation Cetus is at the southern corner of an equilateral triangle whose northern corners are marked by the Pleiades and the star Hamal in Cetus. Twenty-three degrees of arc separates each of those objects.

MEANING:

This star represented the "Tip of the Nose" in Ptolemy's description of the Cetus creature. Its present name is a variation of an Arabic translation "The Nostrils" of the earlier Greek designation.

LORE:

Some traditional star maps showed Menkar at the open mouth of the Whale - Sea Monster figure of Cetus. The closed figure of six stars of which Menkar is a part marks "Head of Cetus" in Greek and Roman sky lore. This same rough circle of stars represented a "Hand of the Pleiades" in early Arabian sky legends and also "Heaven's Round Granary" in astronomical lore of ancient China.

VISUAL ASPECTS:

Menkar is a yellow-orange giant star whose luminosity is approximately 500 times that of the Sun. The surface temperature of a star with this type of spectrum is estimated to be 3,800 K.

PHYSICAL NATURE: **The mass of Menkar is about 2.3 times that of the Sun, and as a class M1.5 giant, it has about 89 times the Sun's diameter. With such a size, 700,000 Suns could fit inside. The 2007 reduction of the Hipparchus parallax data indicates Menkar is about 250 light years from Earth.**

MENKAR - ADDITIONAL DATA:

Catalogue Designations: Alpha Ceti, α Cet, 92 Ceti, HR 911, BD +03°419, HD 18884, SAO 110920, HIP 14135

****

Parallax: 13.1 mas
Distance: 76 parsecs, 250 light years

****

Apparent magnitude: 2.5
Absolute magnitude: -1.9
Luminosity: 498 solar
Luminosity class: giant III

****

Apparent color blend: yellow-orange
Peak color: orange-red
Color index (B-V): 1.64
Spectral type: M1.5
Surface temperature: 3,800 K

****

Diameter: 89 solar
Mass: 2.3 solar

****

Coordinates: (J2000.0) 03h02m +04°05' / 45.57 +4.09
Transit at 9 p.m. November 6
Proper motion: -10 -77 mas/year
Radial velocity: -26 km/sec.

****

NEARBY FEATURE:

Mira "the Magnificent" "the Wonderful" long-period variable star Cetus **(2h19m – 2°59') – Mira is the prototype for a class of long-period pulsating variable stars first identified as variable by German astronomer David Fabricius (1564-1617) in 1596. This orange star is 15 angular degrees southwest of Menkar and it has one of the coolest surfaces known for a star, between 1,900 K and 2,400 K.**

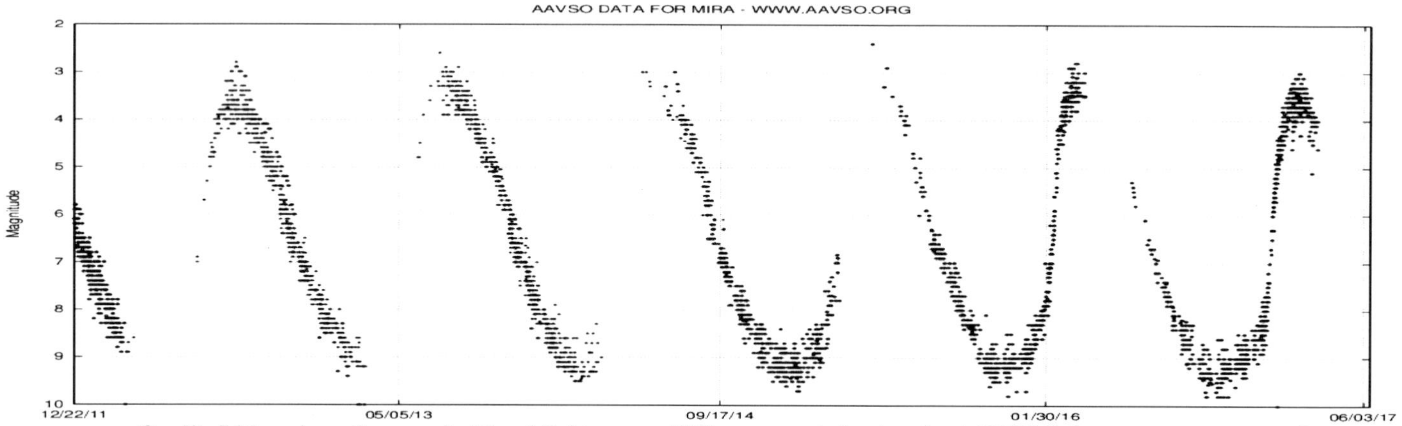

Credit: Lithopsian - Own work: Visual light curve of Mira, generated using the AAVSO light curve generator tool

Mira is a supergiant whose diameter was measured by interferometer to be approximately 420 times that of the Sun. That indicates a volume over 74 million times greater than the Sun! Compared to the Solar System, at its average diameter, Mira would extend outwards to halfway between the asteroid belt and Jupiter's orbit, and when it swells to maximum size, it would nearly reach Jupiter's orbital distance. Using an analogy, if the Sun had the size of a baseball, Mira on average would be about one hundred feet across and cover the infield of a ball park. Alternately, if the Sun was represented by a model the size of a soccer ball, a corresponding Mira model would be 88 meters across, capable of covering nearly an entire pitch.

This great star, which is near the end of its life cycle, varies in brightness between magnitudes of between three or four at maximum, to an invisible to the naked eye magnitude nine at minimum light over a period of about 331 days. With surface temperatures that vary between about 1,900 K and 2,400 K, molecules such as titanium oxide and even water vapor have been detected in the outer atmosphere of Mira.

For twenty years, George Lovi (1940-1993) wrote *Rambling Through the Skies,* the center-fold feature of *Sky and Telescope* magazine. He was our friend, a colleague at the American Museum-Hayden Planetarium and the author of several very well-regarded books on astronomy and celestial cartography. Considering how amateur astronomers can gauge Mira's visual magnitude changes by comparing its brightness to adjacent stars listed on an American Association of Variable Star Observers (AAVSO) reference chart, he noted that in that system, neighboring stars are labeled with two-digit magnitude reference numbers, enabling efficient interpolation when estimating Mira's current brightness status.

Ever since its beginning in 1911, members of the American Association of Variable Star Observers have made important original celestial discoveries as citizen scientists - long before that appellation was even coined. Whether working individually, collaborating with amateur and/or professional colleagues, or outreaching astronomy to other communities, Observers enjoy engaging with the universe and contributing to its understanding.

November Risings

	Prime Season
2 Capella – Auriga	November – April
5 Hassaleh – Auriga	November – April
10 Aldebaran – Taurus	November – March
14 Menkalinan – Auriga	November – April
15 El Nath –Taurus	November – March
19 Mahasim – Auriga	November – April

Capella November 2

North-Northeast Northeast East-Northeast

Auriga

November – April

kah-PELL-ah

0.08 mag. (blend), two very close similar white giant stars; 43 ly; spectroscopic binary, orbit period: 104.0 days, separation: 0.72 A.U.

****

If we visualize the Sun being the size of a baseball by comparison using the same scale, the larger component star in the Capella system would be about three feet wide. This model would be about 13,400 miles from the baseball/Sun, separation about equal to a trip halfway around the world.

In comparison to a soccer ball sized Sun model placed on center spot of a playing field, at that scale a metric model of Capella's largest component would be about 2.7 meters wide. Scaled distance from the soccer ball/Sun to this Capella model would be about 64,300 kilometers, equivalent to a journey one and a half times around Earth.

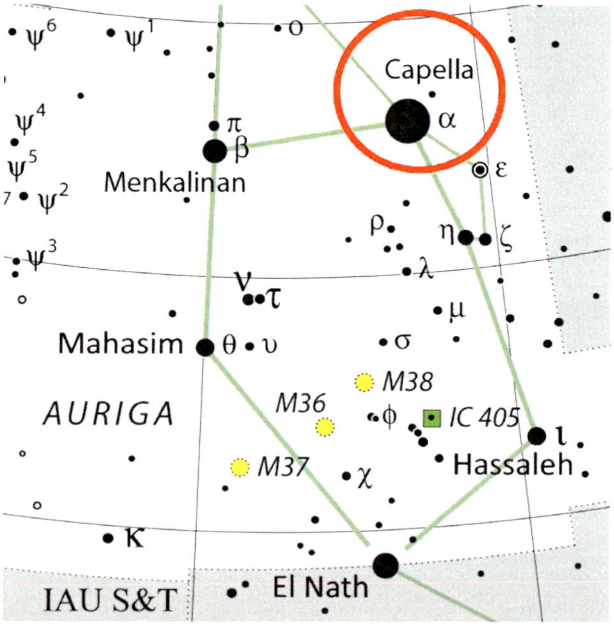

LOCATION:

Capella is a brilliant star in the constellation Auriga, the Charioteer. It is about 20 degrees east of Mirfak.

MEANING:

The name Capella comes from ancient Rome where it was characterized as the "Goat Star".

LORE:

In Mesopotamia, Capella was known as the "Leader" because it was the brightest star in the vicinity of the Sun at time of the Vernal Equinox, then considered as beginning of a new year. In sky lore of Greece and Rome, Capella was often associated with rain and storms; when this star was seen in predawn skies of springtime. Capella was also associated with a "Cornucopia" or "Broken Goat's Horn. Ptolemy described this star as marking Auriga's left shoulder.

HISTORY:

In 1899, spectroscopic analysis of Capella performed by William Wallace Campbell (1862-1938) at Lick Observatory showed that it is a spectroscopic binary. During 1920, John A. Anderson (1876-1959) used an interferometer on the 100-inch telescope at Mount Wilson Observatory, then the world's largest, to help determine orbital characteristics of the Capella binary system. He indicated a separation of about 0.05 arcsecond, revealed by this first use of an interferometer for orbit determinations.

VISUAL ASPECTS:

Diamond like brilliance of Capella is the blended light from a pair of nearly identical giant stars. This binary system is the fourth brightest stellar object in our night sky. With an apparent combined magnitude of 0.1, this magnificent system is more than twice as bright as first-magnitude stars such as Spica and Antares and its glory dominates the northern part of the evening sky from autumn through winter well into springtime. Due to high northern declination, Capella is visible

at some time during the night throughout the year for observers at mid-northern latitudes. Yellow is the strongest individual color in each component's spectrum, and the visual mix of all Capella's colors appears white to our eyes.

PHYSICAL NATURE:

The two giant components orbit their common center of mass at a separation approximately equal to the distance between the Sun and Venus in our Solar System. Each has an orbital period of just over 104 days. The slightly larger component Capella Aa has mass, determined from orbital calculations, 2.7 times that of the Sun. This star has white apparent color and is 74 times more luminous than the Sun. Its surface temperature is estimated at 4,900 K, and its diameter is approximately 12.2 times larger than the Sun. Capella Aa is 1,800 times more capacious than our nearest star.

The second component (Ab) of these near twins in the Capella system has a mass estimated at 2.6 times that of the Sun. Its diameter is 9.2 solar; with its volume being 780 times larger than the Sun. Component Ab has luminosity approximately 64 times the solar value.

Component Aa/Ab surface temperatures are about 4,900 and 5,700 K. respectively. The less massive member Ab has about 79 percent of its larger neighbor's luminosity, despite having a somewhat greater temperature. Because this hotter star's diameter is considerably less, it provides smaller surface area to radiate light; and the higher temperature of component Ab isn't quite enough to overcome its neighbor's size advantage in luminosity assessment.

A third component, known as Capella H, is a yellow-orange spectroscopic binary with same proper motion as the two giants described above. Components of the "H" double system are dwarf stars with a combined apparent magnitude of about ten and separation from the bright dual stars of Capella of 723 arcseconds. That's an angular span about 40% of the Full Moon's apparent diameter.

CAPELLA - ADDITIONAL DATA:

Catalogue Designations: Aa/Ab, Alpha Aurigae, α Aur, 13 Aurigae, HD 34029, HIP 24608, HR 1708, SAO 40186
****

Parallax: 76.2 mas
Distance: 13.1 parsecs, 43 light years
****
Data for Primary Components Aa / Ab:
****
Apparent magnitudes: 0.08 (0.76 / 0.91)
Absolute magnitudes: 0.16 / 0.31
Luminosities: 74 / 64 solar
****

Luminosity classes: giant III / giant III
****
Apparent color blend: white / white
Peak colors: yellow / yellow
Color index (B-V): 0.87 / 0.74
Spectral types: G8 / G1
Surface temperatures: 4,900 K / 5,700 K
****

Diameters: 12.2 / 9.2 solar
Masses: 2.7 / 2.6 solar
****
Coordinates: (J2000.0) 05h17m +46°00' / 79.17 +46.00
Transit at 9 p.m. January 25
Proper motion: +75 -427 mas/year
Radial velocity: +29 km/sec.

NOTE:

A while back, while reading Dickens' *Pickwick Papers,* savoring those remarkable word images of mid-19th century England we noted these lines, at the start of chapter 20, embedded in vivid descriptions of Victorian goings on:

"In the ground-floor front of a dingy house, at the very farthest end of Freeman's Court, Cornhill, sat the four clerks of Messrs. Dodson & Fogg, two of his Majesty's attorneys of the courts of King's Bench and Common Pleas at Westminster, and solicitors of the High Court of Chancery—the foresaid clerks catching as favourable glimpses of heaven's light and heaven's sun, in the course of their daily labours, as a man might hope to do, were he placed at the bottom of a reasonably deep well; and without the opportunity of perceiving the stars in the day-time, which the latter secluded situation affords."

The supposed possibility of seeing stars with naked eyes from the bottom of a deep well or tall chimney in broad daylight is an ancient myth, heard at least since the time of Aristotle. However, a clear daytime sky is far too bright for observing with unaided eyes any star other than the Sun, even if such an object passes through the zenith, which Capella does for observers at latitude 46° North, the same angle as its declination.

To get a feel for the impossibility of naked eye deep well daylight star gazing, consider this – Sirius, brightest star in the night sky, is about 18 times fainter than Venus at its most vivid phase. Even when that planet is at its brightest, it's still a challenge to spot Venus during daytime without optical help.

On the other hand, with a telescope, observing bright stars during daytime is a different story. John Flamsteed, first Astronomer Royal of England, in 1679 was studying Eltanin as part of his unsuccessful efforts to determine the distance of that star. He worked during twilight, using a telescope from the bottom of a deep well that served as its tube. He and his eyepiece were at the bottom of the well, with the objective lens mounted at the top! Observing and measuring positions of stars faint as second and third magnitude during daytime was routinely accomplished in broad daylight, without supposed help of a well, by James Bradley who reported such observations during his own stellar parallax distance hunts in the 1720's.

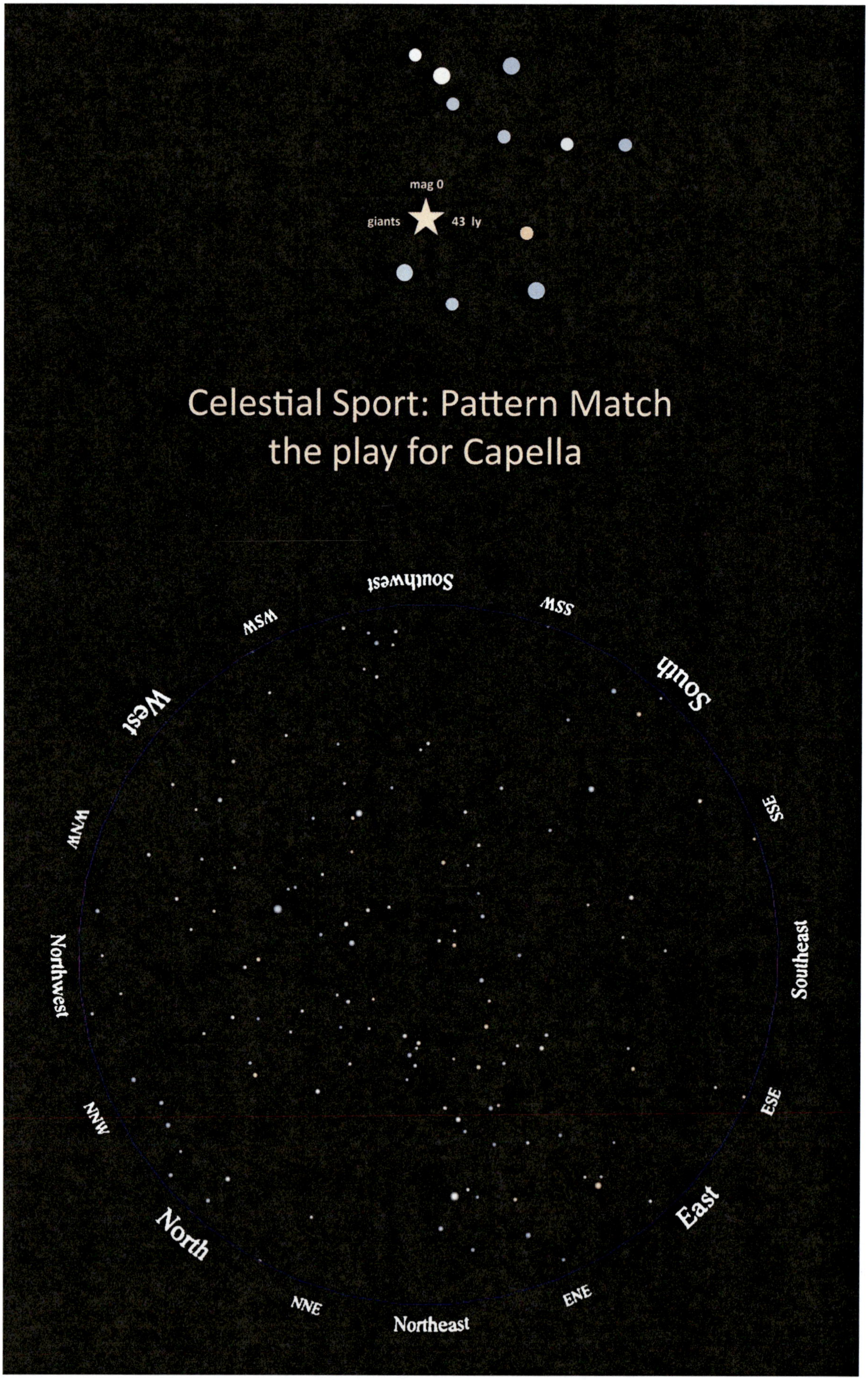
mag 0
giants
43 ly
Celestial Sport: Pattern Match
the play for Capella
Southwest
SSW
South
SSE
Southeast
ESE
East
ENE
Northeast
NNE
North
NNW
Northwest
WNW
West
WSW

Auriga, the Charioteer

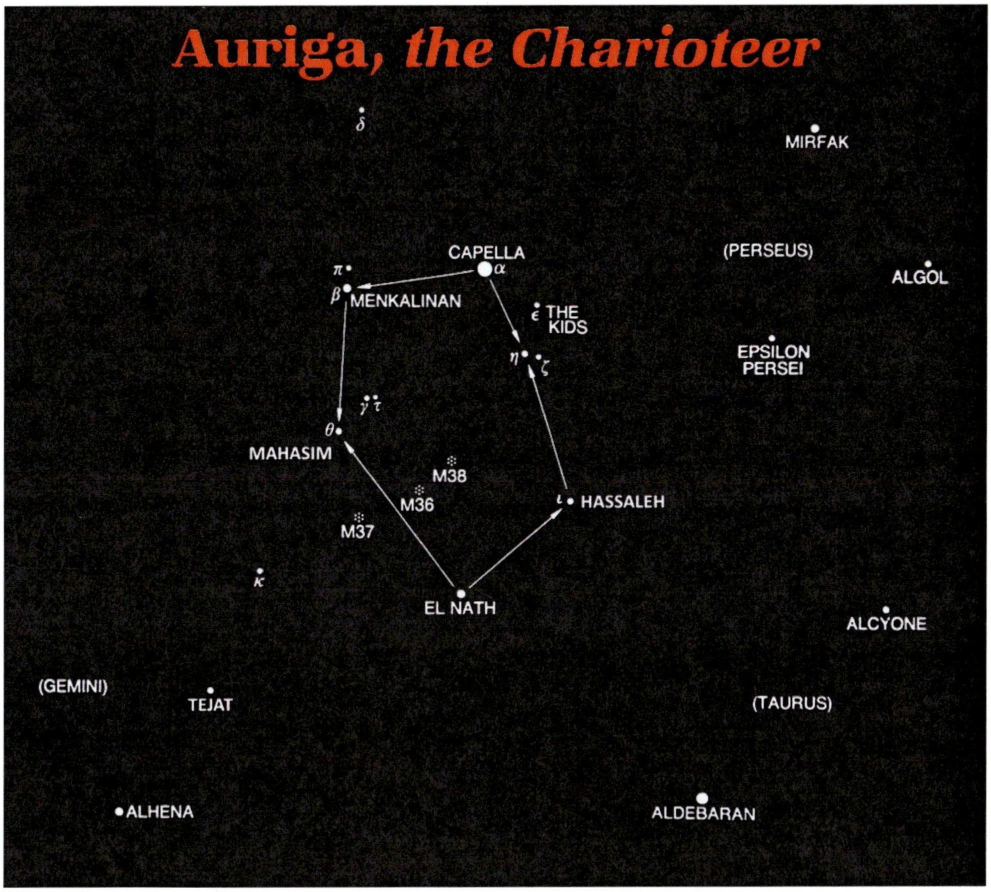

aw-RYE-guh

Aur – Aurigae, *aw-RYE-ghee*

>>>>>>>>>>>>>>>>>>>>>>

Brightest Stars:

November 2 Capella 0.1 (combined) two white giants 43 ly

November 5 Hassaleh 2.7 yellow giant 490 ly

November 14 Menkalinan 1.9 blue-white sub-giant 81 ly

November 19 Mahasim 2.6 blue-white peculiar 170 ly

>>>>>>>>>>>>>>>>>>>>>>

EVENING SEASON:

November - April

SKY TRACK:

Auriga rises above the northeastern horizon, passes overhead, and sets toward the northwest.

LORE:

Characterization of Auriga's stars as a "Charioteer" probably originated in the ancient Middle East. In Greek legends, Auriga was associated with Erichthonius, son of Hephaestus the god of fire and "Mother Earth". Erichthonius was a mythical early king of Athens who was lame and was said to have invented the chariot as a means of transportation.

VISUAL ASPECTS:

Auriga extends across the center of the Milky Way, the Galactic equator line and it contains several open star clusters visible in small telescopes. Three stars near Capella, Eta, Epsilon, and Zeta Aurigae form a small triangle collectively known as the "Kids". El Nath, the star at the southeastern comer of the "Pentagon" asterism of Auriga, is formally assigned to Taurus; thereby serving to mark one of the Bull's horn tips.

NEARBY FEATURE:

Almaaz (Epsilon Aurigae) unusual eclipsing binary star system **(5h 02m +43° 49') – Almaaz (ε Aur) is a third-magnitude object just three degrees**

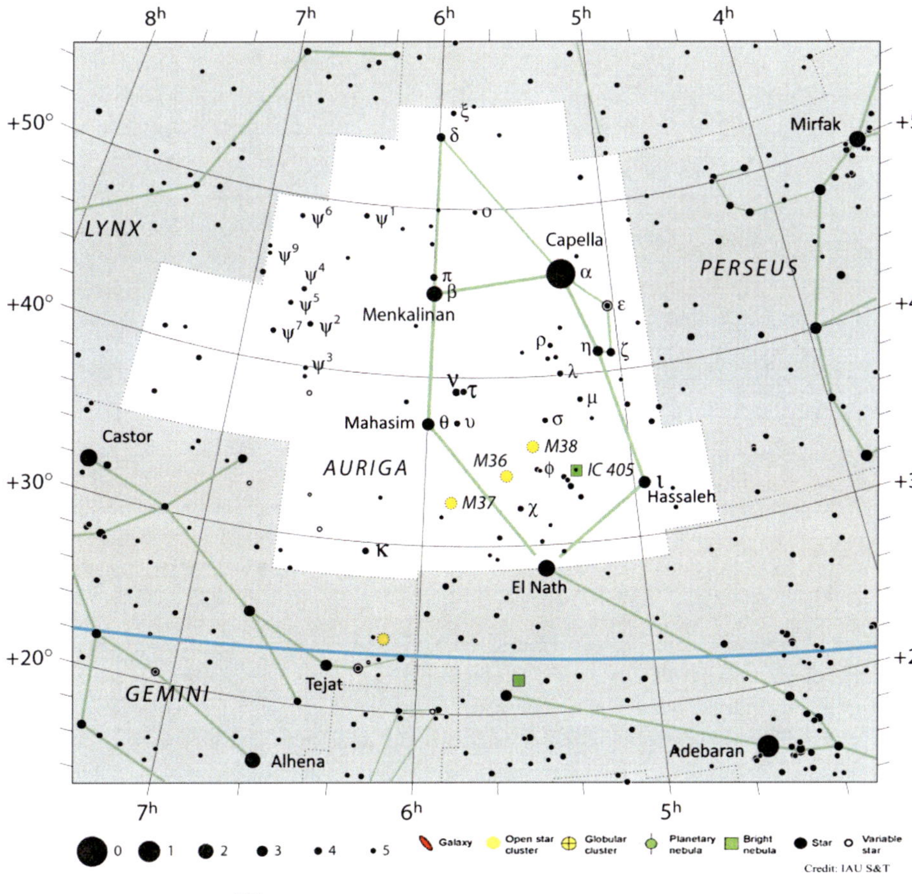

southwest of Capella. It's at the northern tip of the three forming the "Kids of Auriga". Almaaz is a blue-white supergiant estimated to be 2,100 light years from Earth. The star's brightness, diameter, and spectrum vary slightly over a period of about 110 days.

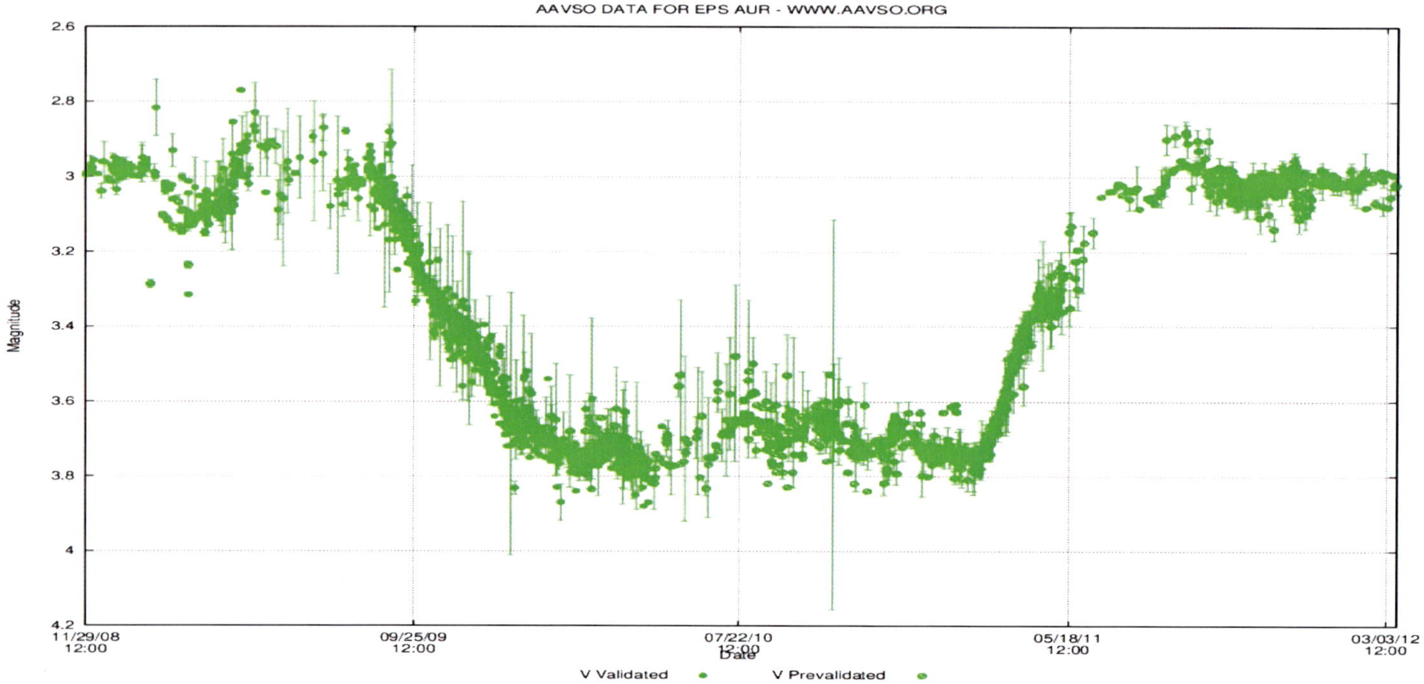

Light Curve of Almaaz from AAVSO Data - Credit: Lithopsian

Auriga – Bayer Uranometria 1603

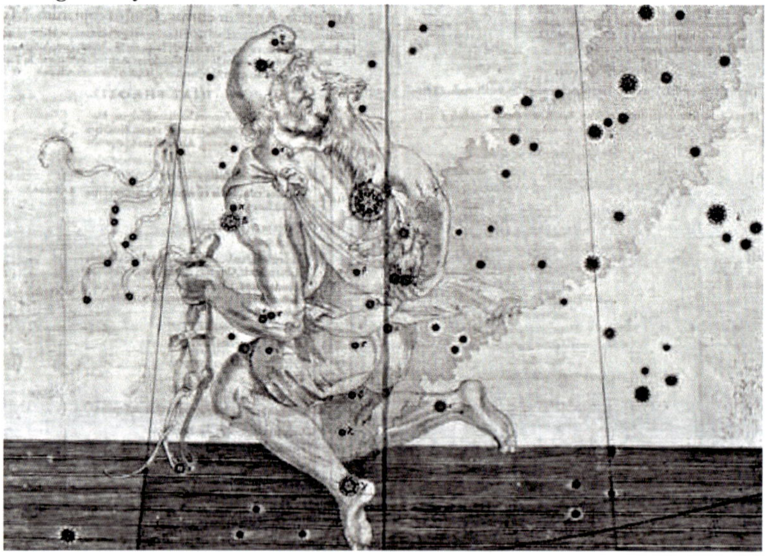

The remarkable characteristic of Epsilon Aurigae which causes it to be especially interesting to astronomers is a significant eclipse that reduces the primary star's apparent magnitude from 2.9 to 3.8 about once every 27 years. Something slowly moving along a large orbit glides into position in front of Epsilon, thereby reducing flux of light reaching Earth to about half the star's normal brightness. It requires about a half year for the light to fade then minimum brightness is maintained for about one year. Restoration of the star to magnitude 2.9 requires an additional half-year. The extraordinary duration of the eclipse indicates the eclipsing object is unusually large!

Astronomers have proposed various models to explain the lengthy eclipse of Epsilon Aurigae, which last occurred in 2009. For many decades the standard explanation has been the eclipsing object is a gigantic star, perhaps 3,000 times the diameter of the Sun. Searches during the 1980's using infrared telescopes however failed to detect infrared radiation from Epsilon Aurigae's companion, and the super-supergiant theory was put aside as alternative theories were considered. In 2009 during the International Year of Astronomy citizen scientists were alerted to join professionals in focusing on the unusual eclipse of Epsilon Aurigae.

During that year, infrared observations of this strange system using the Spitzer Space Telescope indicated the primary star visible to the unaided eye is a dying supergiant star of spectral type F0. There remain various hypotheses as to the nature of the eclipsing object. One concept suggests it is a spectral type B star embedded in a diffuse disk composed of orbiting gravel sized particles. Gaia mission data indicates this remarkable multi-object system is about 1,350 light years (give or take 300) from us.The International Astronomical Union's *Working Group on Star Names* (WGSN) lists *Almaaz* as the official proper name for Epsilon Aurigae. It was seen as the star's name in writings of Ulug Beg.

Hassaleh November 5

Northeast **East-Northeast** **East**

Auriga

November – April

HASS-ah-lay

2.7 mag. yellow giant; 490 ly

Anchor Point Star K3 II

****

Try to imagine the Sun being the size of a baseball. On that scale, Hassaleh would be about 31 feet wide – one and three-quarter times the width of a pitcher's mound. Its proportionate distance from the baseball/Sun would be about 152,600 miles, the same as travelling six times around Earth or two-thirds of the way to the Moon.

If the Sun had the diameter of a Soccer ball, a Hassaleh model at this scale, would be 28 meters wide – one and a half times the size of a playing field's center circle. Scaled distance from the soccer ball/Sun to Hassaleh's model would be about 732,600 kilometers, a span nearly twice the Moon's distance from Earth.

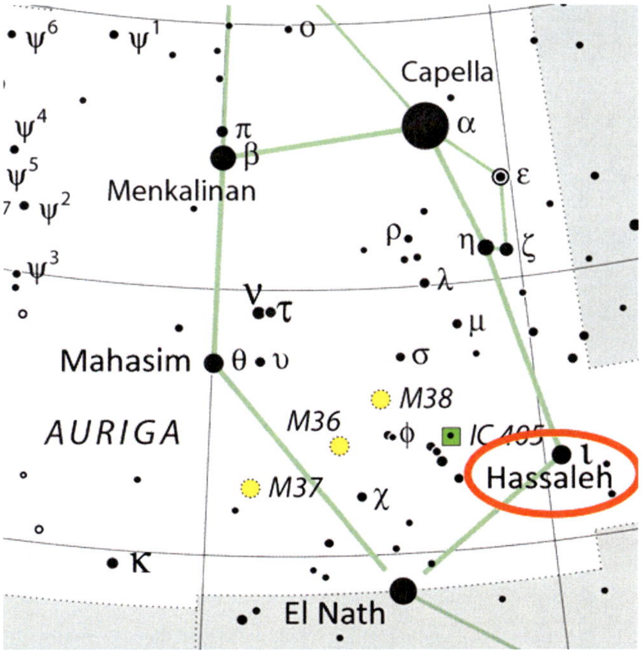

LOCATION:

Hassaleh is at the southwestern corner of a pentagon-shaped asterism outlined by bright stars of Auriga. Shining at third magnitude, it is 12 degrees south of Capella.

LORE:

In Arabian mythology Hassaleh represented "Shoulder of the Rein-Holder", Auriga. It is said that in China it was referenced as "First Star of the Five Chariots".

Use of the name Hassaleh apparently was introduced by the popular mid-twentieth century *Atlas Coeli* produced by Antonín Bečvář, founder of Skalnaté Pleso Observatory in Slovakia.

Etymologist Paul Kunitzsch cited no other source for this name.

VISUAL ASPECTS:

Hassaleh has an apparent magnitude of 2.7. It's a yellow giant at a distance estimated to be about 490 light years from Earth. Stars of this spectral type and luminosity class are about 1,600 times more luminous than our Sun and have a surface temperature of about 4,200 K. Absorption lines of neutral metals are the strongest seen in the spectrum of this star.

Annual proper motion is seven milliarcseconds eastward and 15 milliarcseconds to the south. The combined proper motion is 17 mas toward the southeast. At that rate, Hassaleh would take about 109,000 years to traverse a distance on the celestial sphere equal to the apparent diameter of the full Moon. The star's radial velocity, revealed by Doppler shift measurements of its spectral lines, is +18 kilometers per second. The plus sign indicates that the distance between Hassaleh and the Sun is increasing.

PHYSICAL NATURE:

All spectral type K3 luminosity class II bright giant stars are categorized as such because their spectra appear similar to that of Hassaleh, one of the defining Anchor Point Stars introduced with the MKK stellar spectrum classification system in 1943.

33. K3

The spectral type is determined from the ratios λ 4226: λ 4325 and λ 4290: λ 4299. Luminosity classes are determined from the ratios λ 4071: λ 4077, λ 4063: λ 4077, λ 4045: λ 4077, λ 4260: λ 4215 and λ 4325: λ 4340.

TABLE 28
STANDARDS AT K3

Star	MKK	α	δ	m	HD
δ And	K3 III-IV	00:34	+30°19'	3.5	K2
ι Aur	K3 II	04:50	+33 00	2.9	K2
α Hya	K3 III	09:22	-08 14	2.2	K2
ρ Boo	K3 III	14:27	+30 49	3.8	K0
ε CrB	K3 III	15:53	+27 10	4.2	K0
π Her	K3 II	17:11	+36 55	3.4	K5
λ Her	K3 III	17:26	+26 11	4.5	K0
α Sct	K3 III	18:29	-08 19	4.1	K0
1 Lac	K3 III	22.11	+37 15	4.2	K0

From: ***AN ATLAS OF STELLAR SPECTRA WITH AN OUTLINE OF SPECTRAL CLASSIFICATION*** **Morgan * Keenan * Kellman**
Credit: Astrophys. monographs, Univ. Chicago Press (1943)

The chart above is excerpted from an interactive version of the classic 1943 MKK Atlas, as prepared by Caltech. For over 80 years versions of this venerable atlas have been a basis for the stellar classification system. The list shows Iota Aur (Hassaleh) as one of two stars, along with Pi Her, whose spectrum's appearance defines the standard for spectral type K3 and luminosity class II (Bright Giant stars).

The following associated MKK graphic features comparison photographs from the original MKK Atlas illustrating how very subtle differences in the relative vividness of particular spectrum absorption lines form the basis for classifying a particular star as a Supergiant (class I), a Bright Giant (class II), or as an "ordinary" Giant (luminosity class III).

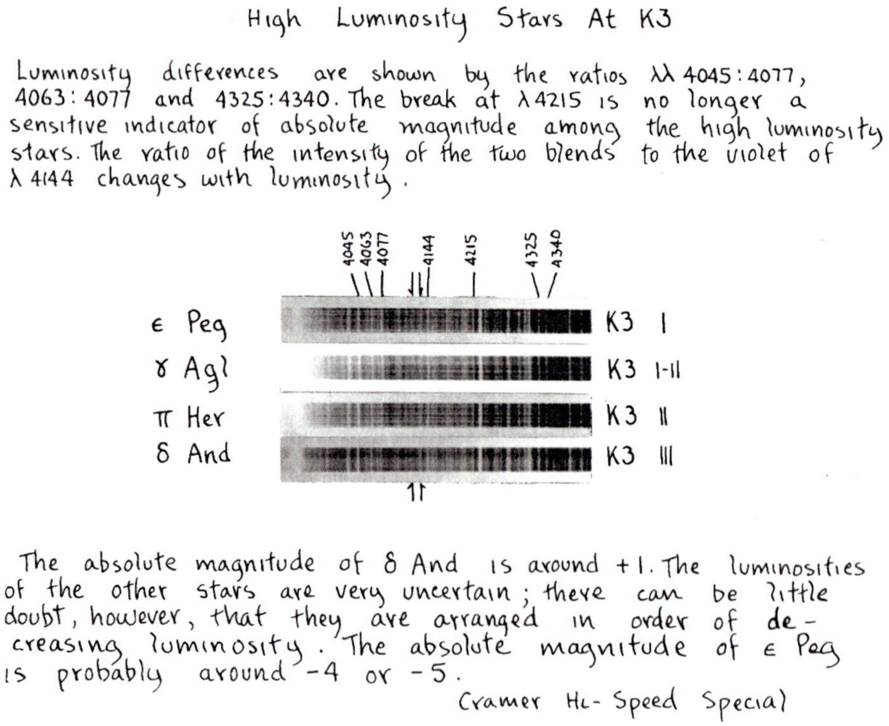

Hassaleh is estimated to have 7.1 times the Sun's mass and 127 times the Sun's diameter. This giant's volume therefore is about 2,050,000 times larger than the Sun.

HASSALEH - ADDITIONAL DATA:

Catalogue Designations: Iota Aur, ι Aur, 3 Aurigae, HR 1577, HD 31398, HIP 23015, SAO 57522, A1105
****
Luminosity class: giant IIv
****
Apparent color blend: yellow
Peak color: orange
Color index (B-V): 1.53
Spectral type: K3
Surface temperature: 4,200 K
****
Diameter: 127 solar
Mass: 7.1 solar
****
Coordinates: (J2000.0) 04h57m +33°10m / 60.95 +33.17
Transit at 9 p.m. January 20
Proper motion: +7 -15 mas/year
Radial velocity: +18 km/sec
****

Parallax: 6.6 mas
Distance: 151.5 parsecs, 490 light years
****

Apparent magnitude: 2.7
Absolute magnitude: -3.2
Luminosity: 1,600 solar

Aldebaran November 10

East-Northeast East East-Southeast

Taurus

November – March

al-DEB-ah-ran

0.86 mag. yellow giant; 67 ly

****

Imagine the Sun as the size of a baseball. At that scale Aldebaran would be about six feet wide – one-third the width of a pitcher's mound. Our stellar model would have to be about 20,900 miles from the baseball/Sun, a distance equal to 84% of Earth's circumference.

If the Sun had the diameter of a Soccer ball, an Aldebaran model, at this scale, would be five and a half meters wide –30% the size of a playing field's center circle. Scaled distance from the soccer ball/Sun to our Aldebaran model would be about 100,200 kilometers, the equivalent of a trip two and a half times around the world.

LOCATION:

Aldebaran continues the rich progression of bright first magnitude stars entering our evening sky at this time of the year. Representing the "eye" of Taurus, the Zodiac's Bull, it is 16 degrees south of Hassaleh and 14 degrees southeast of Alcyone.

MEANING:

The origin of Aldebaran's name is uncertain. It may have been taken from an Arabian description meaning "Driver" or "Follower of the Pleiades".

LORE:

Aldebaran is the ninth-brightest star in our stellar array visible from mid-northern latitudes. Overall, it's the 14th brightest seen from anywhere on Earth (excepting the Sun). This centerpiece of Taurus was celebrated in ancient times as one of four "Royal Stars", which in their turn highlighted each season's evening sky. Others in that set were Regulus heralding spring, Antares of summer evenings, and Fomalhaut, which we like to call The Autumn Star.

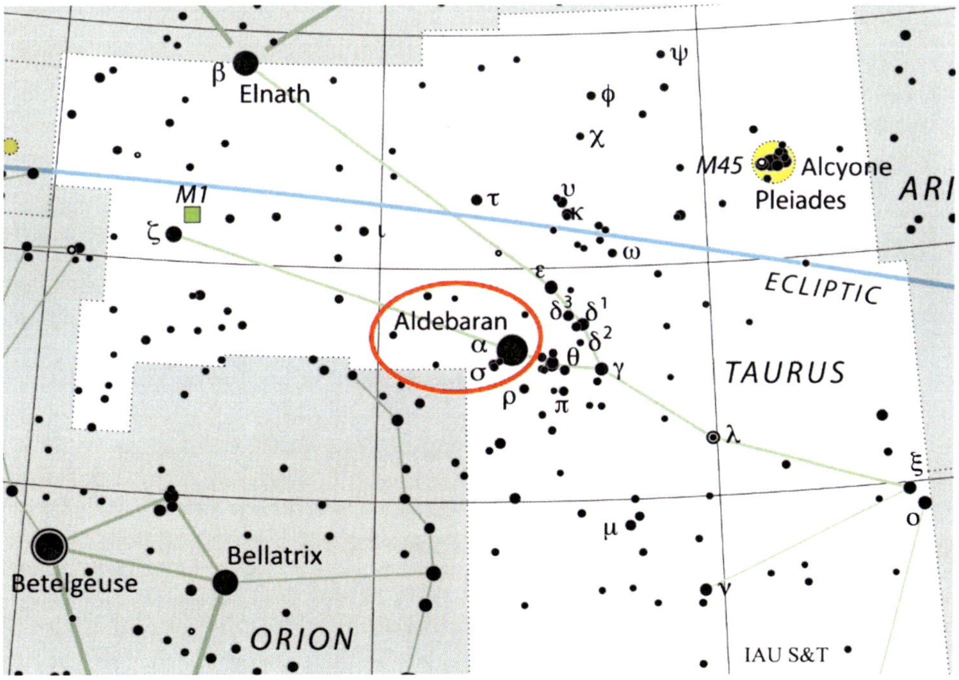

Taurus once contained the sky point of the Vernal Equinox which marks start of northern hemisphere's spring season and in some traditions launching of a new year, during a twenty-two-century era centered around 3,000 B.C. That juxtaposition apparently led to Aldebaran being known as "Leading Star of Stars". In Mesopotamia it was also called "The Messenger of Light", a title applied at various times to Capella and Hamal, as lunar-solar gravitational induced precession of Earth's rotation axis advanced the Equinox site westward over centuries.

Ptolemy described Aldebaran as the "Bright Red Star" in the Hyades, visualizing it as representing an eye of the celestial Bull. His description as red was perhaps a poetic exaggeration prompted by Aldebaran's obvious color contrasts with most bright stars in its neighborhood. The most popular description for the Alpha of Taurus is "Eye of the Bull". The animal's face is portrayed by the distinctive "V" asterism formed by Aldebaran and apparently nearby Hyades cluster stars – adjacent in two dimensions but not three.

HISTORY:

According to Robert Burnham Jr. (1931 – 1993) Edmund Halley, while perusing records of a lunar occultation of Aldebaran observed in Athens early in the sixth century B.C., noted that this star had clearly changed position in celestial latitude and longitude by the time of his own seventeenth century observations. On an expedition to St. Helena island Halley measured coordinates of many stars including Sirius, Arcturus, and Aldebaran (which he called "Palilicium" or "the Bull's Eye").

He wrote: "All these three Stars are found to be above half a degree more *Southerly* at this time (1677) than the Antients (sic) reckoned them … It is scarce credible that the Antients could be deceived in so plain a matter."

By means of Halley's observations, from that island which later became Napoléon Bonaparte's (1769 – 1821) last home, long held notions that stars are eternally fixed in longitude and latitude began yielding to the concept of proper motions – stars having long-term intrinsic movements through space.

In 1803 Napoleon sold French claims to about 828,000 square miles (2,140,000 km^2) of North America to the United States during the administration of Thomas Jefferson. With the stroke of his pen, the Louisiana Purchase doubled the nation's area, added vast resources, provided a perspective to the Pacific, and prevented the United States from being forever boxed in between the Atlantic Ocean and the Mississippi River.

During the winter of 1804-05, halfway through the trip west exploring those new lands, the United States Army Corps of Discovery, led by Meriwether Lewis and William Clark, made it's first cold-weather encampment at a site by two Mandan Indian villages on the upper Missouri River, near present-day Washburn, North Dakota.

At noon on Sunday November 11th, 1804, after "a clear and pleasant morning", Lewis reported his sextant's "altitude" measurement of the Sun's upper limb as 51° 4' 52" (apparently indicated as twice the usual definition of altitude as the angular height of an object above the horizon). From that observation Lewis calculated the Expedition's latitude at the Mandan Villages to be 47° 21' 32.8" N.

That was also the day when a Frenchman, who later would be an interpreter for the Expedition, arrived at Ft. Mandan. It was Toussaint Charbonneau who came with Sacagawea, one of his two Shoshone Indian wives. Next to Lewis and Clark, it is she whose name is most remembered over two centuries later for having been a major contributor to that Expedition's success.

VISUAL ASPECTS:

Aldebaran is a giant whose diameter, measured by means of interferometers, is 25 times that of the Sun. That's equivalent to about half the diameter of Mercury's orbit. This star is 67 light years from Earth, and it shines with an average magnitude of 0.86.

Knowledge of Aldebaran's distance and apparent magnitude enables us to calculate its luminosity, 165 times brighter than the Sun, if each was seen by us from the same distance.

Consistent with creative descriptions of this star as the fiery eye of Taurus the Bull, Aldebaran's surface temperature 3,900 K indicates a yellowish apparent color. Infrared observations suggest Aldebaran is surrounded by a relatively cool shell of gas. As is generally seen with stars of this type and giant class, it's slightly variable in brightness, ranging between magnitudes 0.75 and 0.95 over irregular periods.

PHYSICAL NATURE:

With luminosity class III, spectral type K5 and color index 1.54, Aldebaran comprises a mass 1.2 times the solar amount. It has 25 times the Sun's diameter, corresponding to a volume about 15,600 times larger.

Aldebaran's spectral type shows strong ionized calcium lines, as well as noticeable lines of neutral iron. Neutral calcium is also evident.

Spectral line appearances used for assignment of Aldebaran to spectral type K5 include strength ratios of the calcium (Ca I) absorption line at wave length 4226 Ångstrom units to that of the iron (Fe I) line at 4325 Å wavelength; also, strength of the chromium (Cr I) line at 4290 Å compared to the G Band of Carbyne intensity at wavelength 4299 Å;

In the matter of luminosity class assignment to the giant III class category, that's determined by comparing iron (Fe I) 4063 Å line strength to strontium (Sr II) 4077 Å; and iron (Fe I) 4260 Å line intensity relative to CN Band Cyanogen at 4215 Å.

Line strength ratios used for assignment of a K5 star to a particular luminosity class are examined to see where they would nearly match the array of line strengths seen in various luminosity class anchor point stars of spectral type K5. In the case of Aldebaran's K5 III giant category, its anchor point star is Gamma Draconis, Eltanin.

A yellow-orange dwarf star with a magnitude of 13.5 is a binary orbiting companion, separated from Aldebaran by 31 arcseconds.

ALDEBARAN - ADDITIONAL DATA:

Catalogue Designations: Alpha Tauri, α Tau, 87 Tauri, HD 29139, HIP 21421, HR 1457, SAO 94027

****

Parallax: 48.9 mas
Distance: 20.4 parsecs, 67 light years

****

Apparent magnitude: 0.86 (0.75-0.95) v
Absolute magnitude: -0.7
Luminosity: 165 solar
Luminosity class: giant III

****

Apparent color blend: yellow
Peak color: orange
Color index (B-V): 1.54
Spectral type: K5
Surface temperature: 3,900 K

****

Diameter: 25 solar
Mass: 1.2 solar

****

Coordinates: (J2000.0) 04h 36m +16° 31' / 68.98 +16.51
Transit at 9 p.m. January 15
Proper motion: +63 -189 mas/year
Radial velocity: +54 km/sec.

****

EXOPLANET:

An international team led by Artie P. Hatzes (1957-) in August 2015 reported their detection of a planet designated Alpha Tauri b orbiting Aldebaran at an average distance of 1.46 astronomical units. That separation is comparable to the 1.52 A.U. orbital span between Mars and the Sun. Eccentricities of planets Alpha Tauri b and Mars are 0.10 and 0.09 respectively and their orbital periods, 629 days and 687 days are also similar. However Alpha Tauri b has a mass about 6.5 times greater than Jupiter, whereas Jupiter is about 2,900 times more massive than Mars.

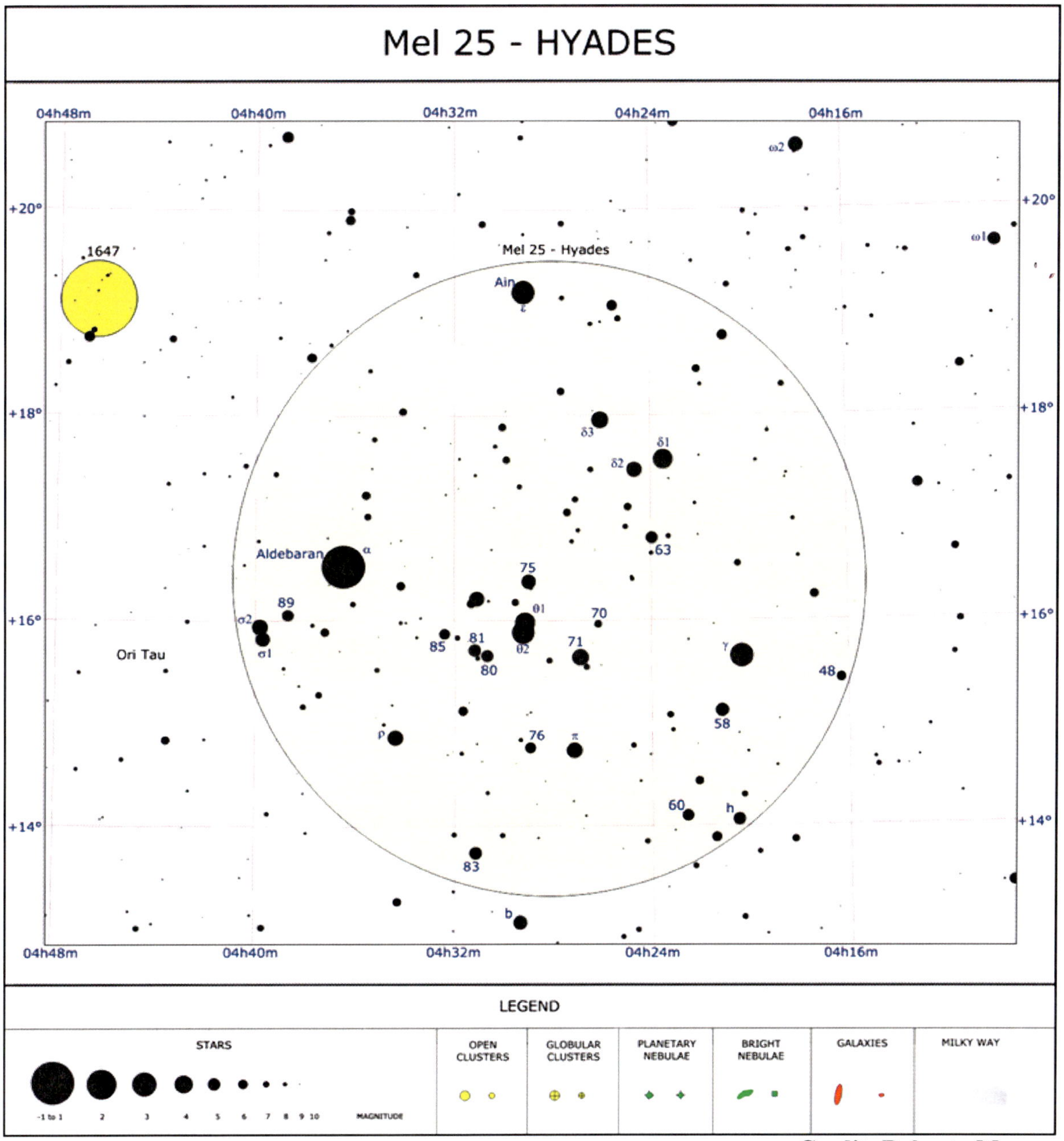

Credit: Roberto Mura

NEARBY FEATURES:

Theta^1 and Theta^2 Tauri – **Binocular pair of stars (4h 29m +15° 58' and 4h 29m +15° 52') – Star Theta^2 Tauri is the brightest member of the Hyades cluster, and it forms a lovely binocular pairing with its physically unrelated apparent neighbor, Theta^1. Their angular separation is 337 arcseconds, about one-fifth the apparent diameter of the Full Moon. Located about two degrees southwest of Aldebaran, Theta^2 is blue-white with apparent magnitude 3.4, while Theta^1 has magnitude 3.8 and is yellow.**

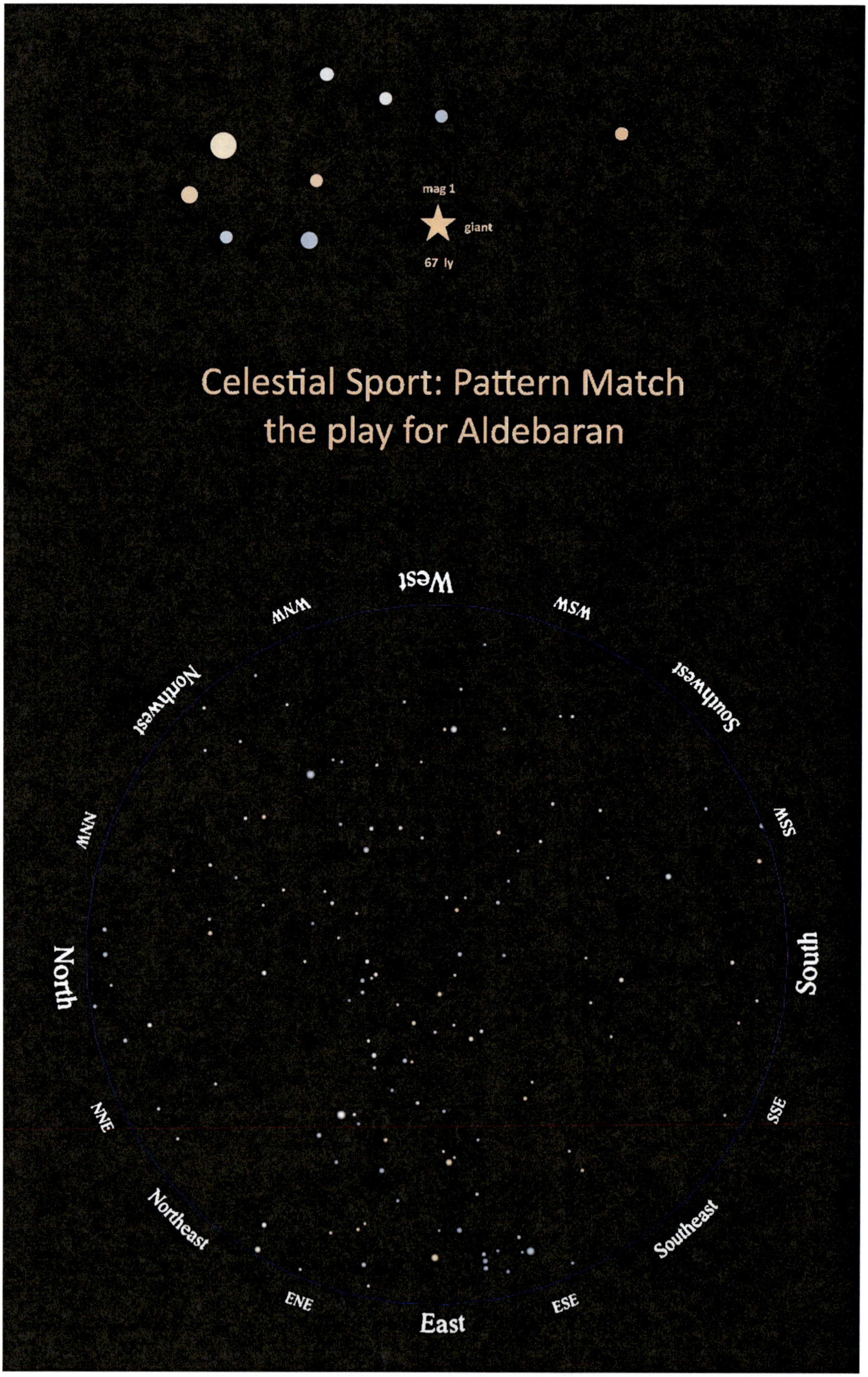
mag 1
giant
67 ly
Celestial Sport: Pattern Match
the play for Aldebaran
West
WSW
WNW
Southwest
Northwest
SSW
NNW
North
South
NNE
SSE
Northeast
Southeast
ENE
ESE
East

Menkalinan November 14

North-Northeast Northeast East-Northeast

Auriga

November – April

men-KAL-ih-nan

1.9 (blend) mag. two blue-white subgiants; 81 ly; eclipsing binary of Algol type, 1.89m-1.98m, period 3.96 days

****

If we visualize the Sun as the size of a baseball, on that scale the two nearly identical components of the Menkalinan binary system would each be about eight inches wide. To be in scale, they would have to be about 25,200 miles from our baseball/Sun model. That's in the ballpark distance of a great circle trip once around the world.

Considering a metric model of the Sun having the diameter of a Soccer ball, models of Menkalinan's components made to scale, would each be 62 centimeters wide. Scaled distance from the soccer ball/Sun to our Menkalinan models would be 121,100 kilometers, equal to a flight three times around Earth's circumference, equivalent to a third of the average distance between Earth and the Moon.

LOCATION:

Menkalinan is a second magnitude star situated at the northeastern (top left when crossing the meridian) corner of the pentagon-shaped asterism of Auriga. It's the second brightest star in this constellation and is positioned seven degrees east of brilliant Capella.

MEANING:

Menkalinan's name was adapted from an Arabian version of the star's Greek description meaning: "Right Shoulder of the Charioteer".

LORE:

The pentagon of Auriga is one of the prime asterism shapes of winter evenings. Look for this stellar figure when the night is clear and dark, helpfully orienting yourself first by recognizing Orion, Sirius, and the Charioteer's brilliant gem Capella. The "Sky Driver" Auriga is shown in traditional images facing us, with the "Kids" tucked by his left arm, with the right free ready for action, driving the chariot, or if challenged wielding his sword.

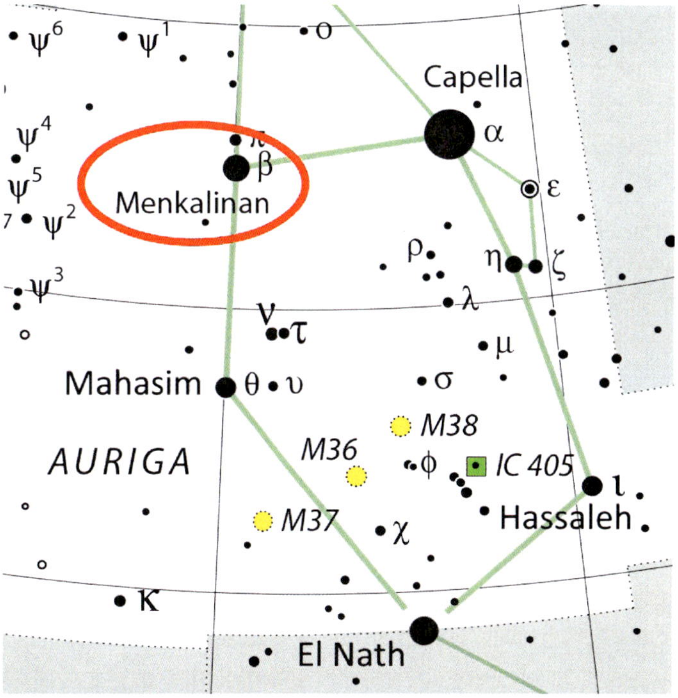

HISTORY:

Antonia C. Maury (1866-1952) of the Harvard College Observatory found during observations she made in 1889 that Menkalinan is a spectroscopic binary.

In 1910 Joel Stebbins (1878-1966), a pioneer in the development of photoelectric photometry, used one of his electric light measuring devices at Washburn Observatory at the University of Wisconsin to study variations in Menkalinan's brightness. His analysis of the star's magnitude changes showed Menkalinan is part of a short period eclipsing binary system. Stebbins was also able to determine the stellar pairs' orbital characteristics from that photometric research.

Some of the "Women Computers" at Harvard College Observatory, who worked for the astronomer Edward Charles Pickering. The group included Harvard astronomers Henrietta Swan Leavitt (1868–1921), Annie Jump Cannon (1863–1941), Williamina Fleming (1857–1911), and Antonia Maury (1866–1952) A chart showing Menkalinan's December 1889 light curve is on the wall .

VISUAL ASPECTS:

The Menkalinan system consists of two virtually identical blue-white subgiant stars in a nearly circular orbit inclined about 12 degrees of arc from our line of sight. Due to this low inclination, we detect the stars as they partially eclipse each other while orbiting. Those partial eclipses cause an eight percent drop in Menkalinan's collective brightness, every three days, 23 hours, and three minutes.

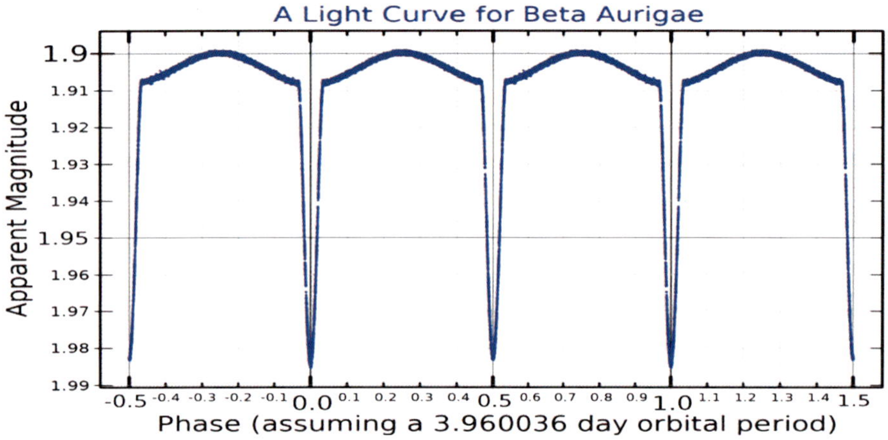

Light Curve of Menkalinan from Southworth et al data - Credit: PopePompus

PHYSICAL NATURE:

Both components of Menkalinan have A1 spectral type and luminosity class subgiant IV classifications. Analysis of their orbit indicates their respective masses are 2.4 and 2.3 times that of the Sun. The components of Menkalinan have diameters 2.8 and 2.6 times our Sun's and luminosities 92 and 43 solar. The volumes of these subgiant stars are 22 and 17.6 times the Sun's size. Respective surface temperatures are 9,350 K and 9,200 K.

The distance of Menkalinan from Earth is revealed by trigonometric parallax to be 81 light years, and its space velocity suggests that it belongs to the Ursa Major moving star group. That's the nearest such stellar assembly to us and it includes five of the Big Dipper's seven stars.

Menkalinan is near the 6-hour line of right ascension, part of the solstitial colure. That great circle intersects the ecliptic at the summer solstice point, the Sun's highest sky position each year for observers north of the Tropics.

MENKALINAN- -ADDITIONAL DATA:

Catalogue Designations: Beta Aurigae, β Aur, (Beta Aurigae A / Beta Aurigae B), 34 Aurigae, HD 40183, HIP 28360, HR 2088, SAO 40750
****

Parallax: 40.2 mas
Distance: 24.9 parsecs, 81 light years
****

Apparent magnitudes: 1.9/2.7
Absolute magnitude: -0.08/0.72
Luminosity: 92/43 solar
Luminosity class: subgiant IV/subgiant IV
****

Apparent color blend: blue-white/blue-white
Peak colors: white/white
Color indices (B-V): 0.03/0.03
Spectral types: A1/A1
Surface temperatures: 9,350 K/9,200 K
****

Diameters: 2.8/2.6 solar
Masses: 2.4/2.3 solar

****

Coordinates: (J2000.0) 06h00m +44°57' / 89.88 + 44.95
Transit at 9 p.m. February 5
Proper motion: -56 -1 mas/year
Radial velocity: -16 km/sec.
****

El Nath November 15

Northeast **East-Northeast** **East**

Taurus

November – March

EL Nath

1.7 mag. blue-white giant; 140 ly

****

Imagining the Sun as the size of a baseball; at that scale El Nath would be a foot in diameter. To maintain the correct proportions an El Nath would have to be placed about 43,600 miles from the baseball/Sun, a distance equal to one and three-quarter times the circumference of Earth.

If the Sun instead had the diameter of a Soccer ball, a metric sized El Nath would be 92 centimeters wide. Scaled distance from the soccer ball/Sun to this model would be about 209,300 kilometers, just over half the Moon's actual distance from Earth.

LOCATION:

El Nath is at the southern tip of the Pentagon of Auriga. It's a second-magnitude star located 17 degrees south of Capella, and nearly the same angular distant from Aldebaran which is southwest of El Nath. For reference remember that an average fist, seen with arm fully extended, is approximately ten angular degrees in length.

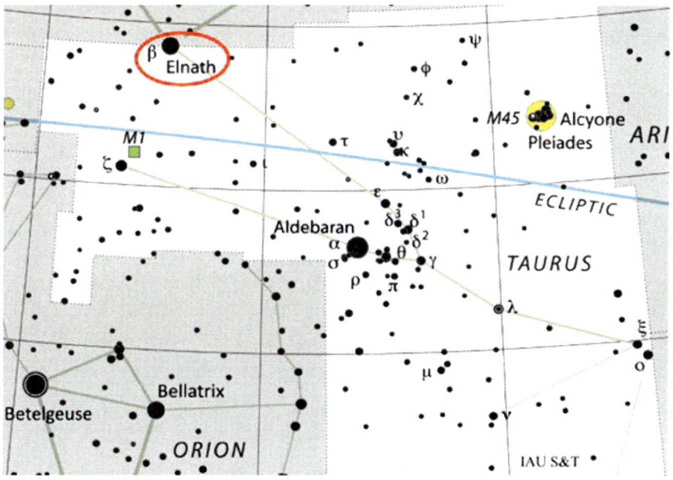

MEANING:

The name El Nath apparently was applied to this star around the beginning of the nineteenth century. It originally had been a name for the star Hamal in Aries and derived from an Arabic description meaning the "Butting One".

LORE:

El Nath marks the northern horn tip of Taurus the Bull. It has a dual identity, as also being traditionally considered part of Auriga, where it represented the Charioteer's heel as well as a point in the pentagon asterism. Although the International Astronomical Union's official constellation borders have El Nath in the realm of the Bull, that doesn't prevent us from visualizing this star as part of Auriga's pentagon, then blinking and again seeing it as the northern horn tip of Taurus.

In ancient India El Nath was associated with the god of fire.

HISTORY:

In 1930, when the International Astronomical Union adopted Eugène Delporte's proposed boundaries defining formal sky zones for all 88 modern constellations, El Nath was placed in the corral of Taurus.

VISUAL ASPECTS:

El Nath is a blue-white giant estimated to be about 140 light years from Earth. Its surface temperature is approximately 13,800 K, and it's 300 times more luminous than the Sun. The space velocity of El Nath suggests that it may be an outlying member of the Pleiades star cluster family.

PHYSICAL NATURE:

With spectrum type B7 and luminosity classification as III giant, El Nath presents a spectrum with strong lines of hydrogen, also notable lines of ionized magnesium and silicon. The star's relatively high temperature, and considerably more extensive surface area, contribute to its luminosity being about 300 times that radiated into space by the Sun.

The mass of El Nath is five times greater than the Sun and its diameter is 4.2 times wider. Therefore, the volume is about 74 times larger than the Sun.

EL NATH - ADDITIONAL DATA:

Catalogue Designations: Beta Tauri, β Tau, 112 Tau, HD 35497, HR 1791, HIP 25428, SAO 77168, A1115

****

Parallax: 24.4 mas
Distance: 41 parsecs, 140 light years

****

Apparent magnitude: 1.7
Absolute magnitude: -1.4
Luminosity: 300 solar
Luminosity class: giant III

****

Apparent color blend: blue-white
Peak color: blue-white
Color index (B-V): -0.13
Spectral type: B7
Surface temperature: 13,800 K

****

Diameter: 4.2 solar
Mass: 5 solar

****

Coordinates: (J2000.0) 5h26m +28°36' / 81.57 +28.61
Transit at 9 p.m. January 27
Proper motion: +23 -174 mas/year
Radial velocity: 9 km/sec.

****

NEARBY FEATURES:

The Crab Nebula, Messier 1 – asi

***M1 Crab Nebula supernova remnant and pulsar Taurus* (5h 35m +22° 01') – On July 4, 1054, Chinese astronomers recorded the sudden appearance of what they called a "guest star" amidst Taurus. The object was as bright as Venus, and for three weeks it could be seen even during daytime. After about twenty-one months it faded to invisibility.**

Although no European records of that object's presence have yet surfaced, petroglyphs etched on rocks in northern Arizona suggest the guest star was noted by Native American observers.

In 1731, English amateur astronomer John Bevis (1695-1771) discovered a wisp of nebulosity, which later would be designated Messier 1, about 6.5 degrees south of El Nath and one degree northwest of third magnitude star Zeta Tauri, which marks the Bull's southern horn tip.

Charles Messier (1730-1817) in 1758 also spotted and recorded the object while concentrating on the Taurus region seeking what would become known as Halley's Comet. Messier assigned it number one on his eventual list of nebulae and star clusters; which he usually considered annoying false positive distractions in his comet hunting quest.

By 1844, William Parsons, 3rd Earl of Rosse (1800-1867) in Ireland, using his mammoth 72-inch reflecting telescope, discovered filaments emanating from M1's center that reminded him of crab legs, thereby inspiring the object's name. The Crab Nebula, M1 appears as a hazy light oval in small telescopes. Images taken with larger instruments are needed to reveal the Crab's elaborate structure.
The Crab extends across about seven minutes of arc, roughly one quarter the angular width of the Moon's disk and its distance is estimated at about 6,500 light years. It's one of the best known and dramatic examples of a supernova remnant.

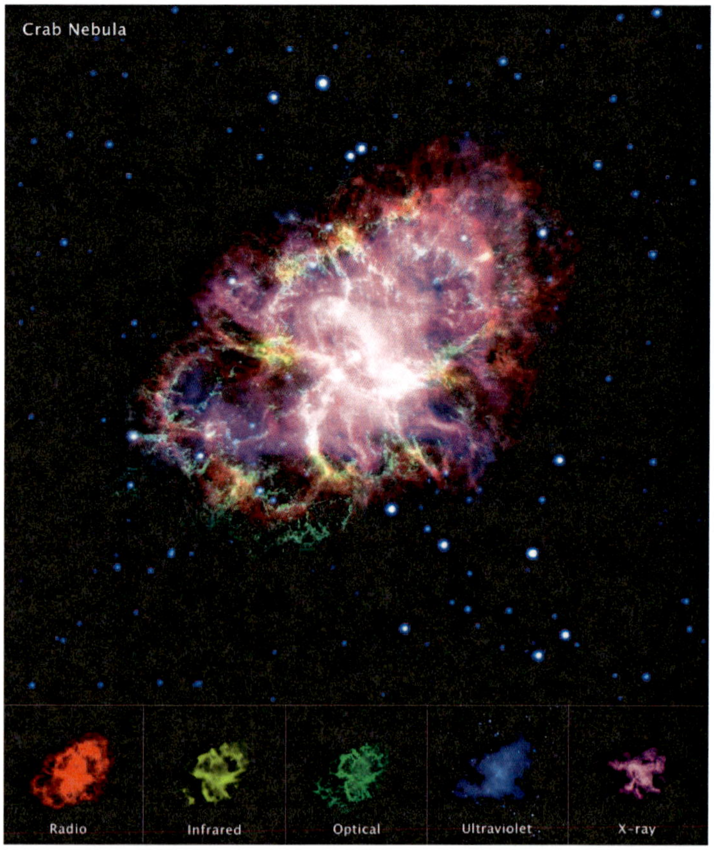

Crab Nebula in five (composite) wavelengths Credit: NASA, ESA

Visible nebula details represent outer layers of a long-gone massive star whose explosive end became apparent on Earth back in 1054. The progenitor supergiant's core had collapsed, becoming a neutron star with a diameter of about 20 kilometers and density equal to that of an atomic nucleus. Such super crushed objects had been postulated since the 1930s by various astronomers as an end-product of massive star explosions at the conclusion of their energy producing terms of existence. Such exceptionally massive stars do not lose sufficient substance by jettisoning gas shells therefore when their nuclear fuel runs out, they still exceed Chandrasekhar's hair-trigger limit of 1.4 solar masses and they can't dodge disaster. At termination of energy production, they collapse with their cores imploding to form neutron stars, black holes, or simply disintegrate, while rebounding of outer stellar layers blast away into surrounding space.

In the case of neutron stars/pulsars, such as that deep in M1, internal core pressures due to gravity are so great, protons and electrons squeeze together forming neutrons.

At the center of the Crab Nebula is the Crab Pulsar, rotating 30.2 times per second while generating electromagnetic radiation across the spectrum ranging from gamma rays to radio waves.

Mahasim November 19

Northeast East-Northeast East

Auriga

November – April

MAH-ha-sim

2.6 mag. blue-white peculiar; 170 ly; variable star of the Alpha² CVn type, double star A0p+)

****

Imagine the Sun as the size of a baseball. At that scale Mahasim would be about 15 inches wide. The model would have to be about 53,000 miles from the baseball/Sun to be at a correctly proportioned separation. That's the same number of miles as a trip two and a quarter times around Earth.

Instead, if the Sun was the diameter of a Soccer ball, a metrically scaled Mahasim model would be one meter and twelve centimeters wide. Equivalent distance from the soccer ball/Sun to our Mahasim model would be 254,200 kilometers, six and a third times Earth's circumference, 66% of the actual average distance between Earth and Moon.

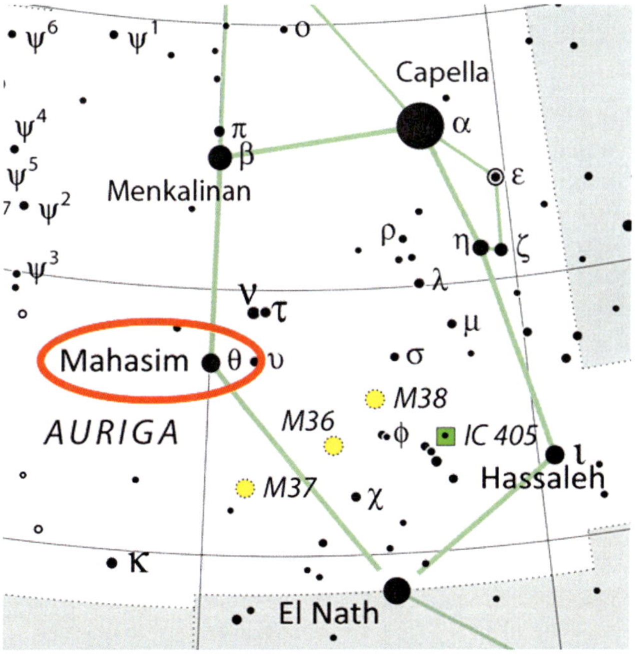

LOCATION:

Third magnitude Mahasim is on the eastern side of the Pentagon of Auriga, seven degrees directly south of Menkalinan.

LORE:

Until 1930, when the International Astronomical Union formalized boundaries, proposed by its Commission 3 led by Eugène Delporte (1882-1955) for the 88 modern constellations, there had been no recognized boundaries between constellations. Prominent stars and groups of stars were regarded as de facto indicators of a constellation's sky domain. Several relatively bright stars were alternately shown as features of two constellations. For example, Alpheratz marks the "Head of Andromeda", Princess of Ethiopia, and also the northeast corner of the Great Square of Pegasus. El Nath, inside the 1930 boundaries of Taurus, spots one of the Bull's horn tips as well as southern corner of Auriga's Pentagon.

The name Mahasim is derived from an Arabic term for the "Wrist" of Auriga, the Charioteer.

VISUAL ASPECTS:

Mahasim has an apparent magnitude of 2.6 and shines with an apparent blue white color. At a distance of 170 light years from Earth, it's is estimated to have luminosity about 184 times that of the Sun. The star, sometimes indicated as component "A" has a seventh magnitude white main sequence companion at an angular separation of 3.9 arcseconds.

PHYSICAL NATURE:

Primary member of the Mahasim system is an A0 spectral type star with surface temperature about 10,400 K. It is categorized as "peculiar" because its spectrum defies the usual appearance of a type A0 main sequence star fusing hydrogen in its core. Mahasim contains 3.4 solar masses in a volume 133 times larger than the Sun. Its diameter is 5.1 times the solar width.

Mahasim's spectrum is peculiar due to unusually prominent absorption lines of the element silicon; causing it to be labeled silicon star. Cor Caroli aka Alpha^2 Canum Venaticorum is the prototype for objects with similar spectral characteristics.

In addition to an unusually strong magnetic field, Mahasim's ionized calcium Ca II K line is extremely faint, but its selenium Sl II line blend at 4128-30Å is strong. Lines of chromium Cr II vary in strength. The MKK Atlas description notes Mahasim appears to be of luminosity class III, brighter than a typical main sequence star of type A0. That Atlas' absolute magnitude range estimate for this star, -1 to -2, is close to recent estimates of -0.9.

MAHASIM - ADDITIONAL DATA:

Catalogue Designations: Theta Aurigae, θ Aur, 37 Aurigae, HD 40312, HIP 28380, HR 2095, SAO 58636, A1119
****

Parallax: 19.7 mas
Distance: 50.8 parsecs, 170 light years
****
Apparent magnitude: 2.6
Absolute magnitude: -0.9
Luminosity: 184 solar
Luminosity class: pSi "peculiar/silicon"
****

Apparent color blend: blue-white
Peak color: white
Color index (B-V): -0.08
Spectral type: A0
Surface temperature: 10,400 K
****
Diameter: 5.1 solar
Mass: 3.4 solar

****
Coordinates: (J2000.0) 06h00m +37°13' / 90 +37.22
Transit at 9 p.m. February 5
Proper motion: +44 -74 mas/year
Radial velocity: +29 km/sec.
****

NEARBY FEATURES:

M37 open star cluster Auriga (5h 52m +32° 33') – There are three attractive open star clusters in the constellation Auriga: M36, M37, and M38, with M37 considered the finest visually. Resembling an unresolved sixth-magnitude patch when seen with binoculars; Messier 37 is approximately five degrees of arc south of Mahasim. When observed through several telescopes of increasing size, M37 begins to be resolved into a marvelous field of sparkling stars. This cluster is believed to contain at least 400 members.

Messier 36 – Open Star Cluster in Auriga -asi

The somewhat truncated upper left end point of its main sequence indicates M37 formed from a cloud of gas and dust about 500 million years ago. Its distance is estimated at 4,500 light years from the Solar System.

The following is an excerpt from the observing notes of Charles Messier , detailing his initial report of the object we now know as Messier 37. However, this open star cluster was first reported in the astronomical record by Giovanni Batista Hodierna before 1654. Independently discovered by Messier, it was recorded by him as #37 (not comets):

September 2, 1764. #37. R.A. 5h 37m 01s (84° 15' 12") dec.+32d 11' 51"
"Cluster of small stars, little remote from the preceding (M36), above the parallel of Chi Aurigae; the stars are smaller, more close together and enclosing some nebulosity; with an ordinary telescope of 3.5 feet (focal length), one has pain to see the stars.

Admiral William Henry Smyth observing M37 in October 1836, benefiting from 72 years of additional telescope development, recorded the following about M37: "A double star in a cluster in front of Auriga's left shin. A and B, both 10th-magnitude, and pale yellow. A magnificient object, the whole field being strewed as it were with sparkling gold-dust; and the group is resolvable into about 500 stars, from the 10th to the 14th magnitudes, besides the outliers. It was found and fixed by Messier in 1764, who described it as "a mass of small stars, much enveloped in nebulous matter." This nebulous matter, however, yields to my telescope, and resolves into infinitely minute points of lucid light, among the distinct little individuals. It is immediately preceded on the parallel by another small double star."

December Risings

	Prime Season
1 Bellatrix– Orion	December – March
3 Tejat – Gemini	December – May
7 Cursa – Eridanus	December – March
8 Betelgeuse – Orion	December – March
9 Mintaka – Orion	December – March
10 Alnilam – Orion	December – March
12 Alhena – Gemini	December – May
13 Alnitak – Orion	December – March
14 Rigel – Orion	December – March
16 Hatysa – Orion	December – March
17 Castor – Gemini	December – May
21 Pollux – Gemini	December – May
24 Saiph – Orion	December – March

The Bright Stars of Winter

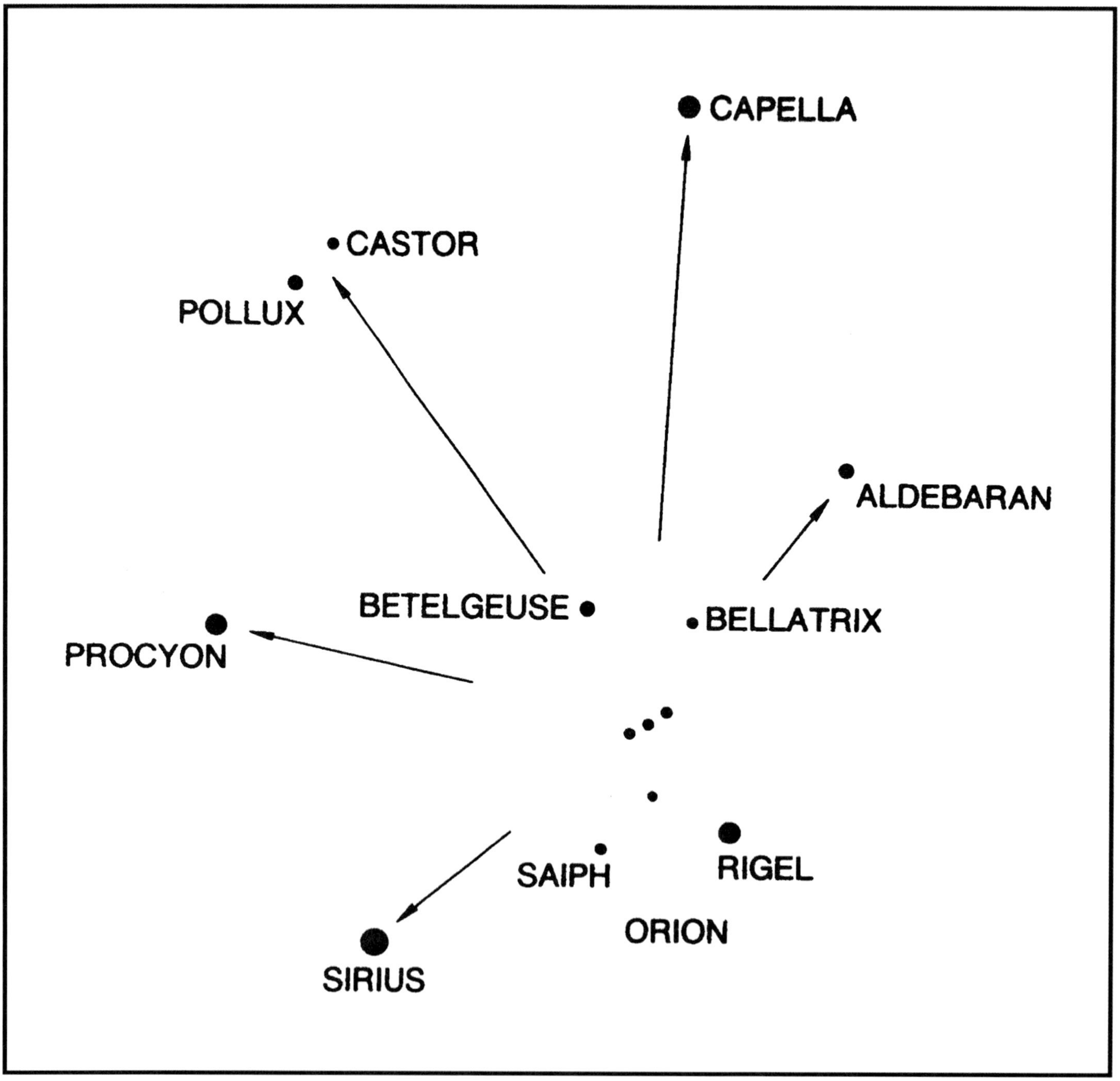

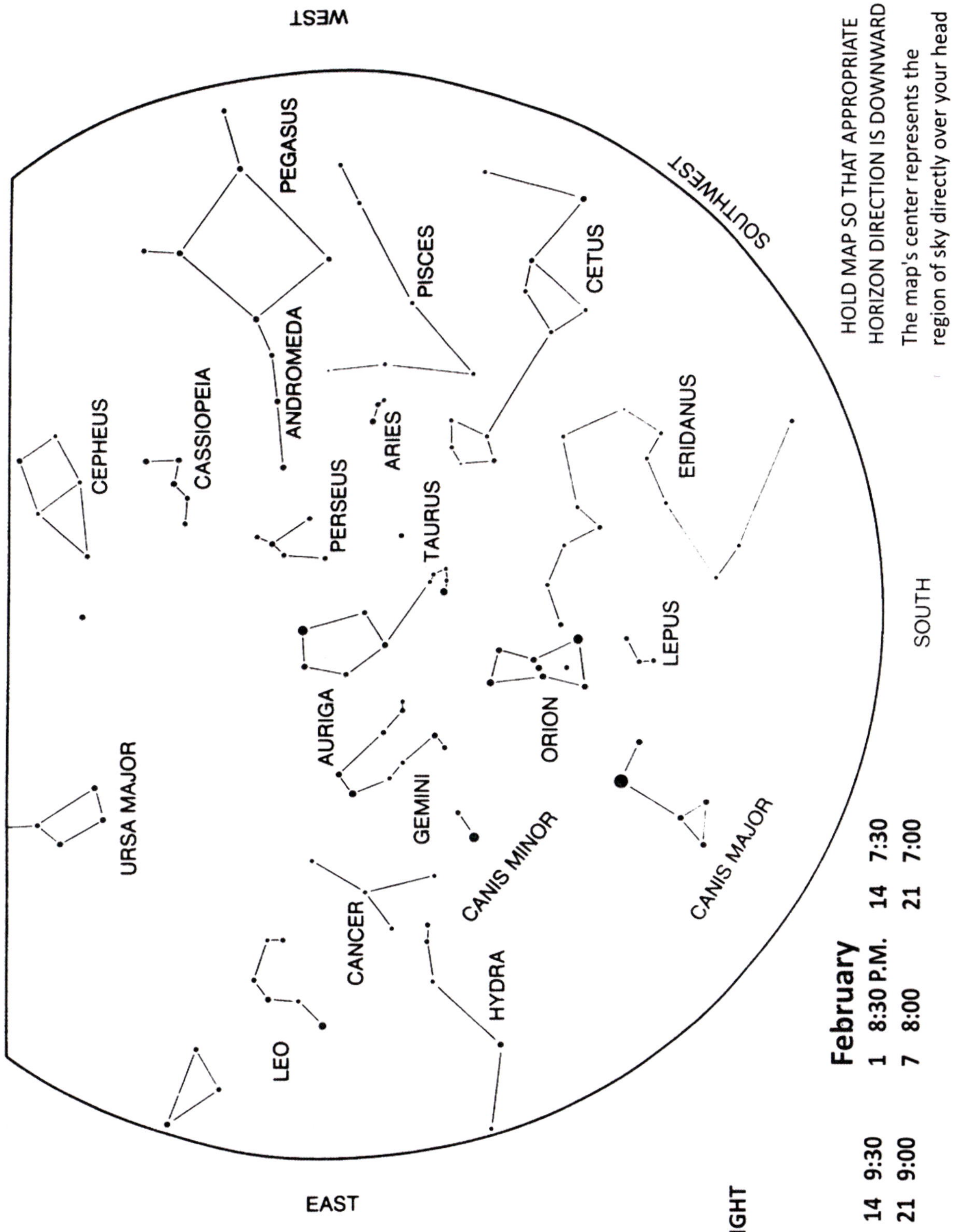

The Winter Sky

December
1 12:30 A.M.
7 12:00 MIDNIGHT
14 11:30 P.M
21 11:00

January
1 10:30 P.M. 14 9:30
7 10:00 21 9:00

February
1 8:30 P.M. 14 7:30
7 8:00 21 7:00

Bellatrix December 1

East **East-Southeast** **Southeast**

Orion

December – March

BELL-ah-trix
1.6 mag. blue-white main sequence; 250 ly
****

If we visualize the Sun being big as a baseball, by comparison Bellatrix would be almost one and a half foot wide. Our model star would then have to be placed at the scaled distance 77,900 miles from the baseball/Sun – that's equal to just a bit more than three trips completely around Earth and 33% the distance to the Moon.

Alternately, compared to a soccer ball sized Sun model placed on center spot of a playing field, a model of Bellatrix would be about one and a third meter in diameter. At its appropriately scaled distance our Bellatrix model would be 373,800 kilometers from the soccer ball/Sun – about equal to the average distance between Earth and Moon.

LOCATION:

Second-magnitude Bellatrix is 16 degrees of arc southeast of Aldebaran. It's the first bright star of magnificent Orion to rise in our eastern sky.

MEANING:

Motivation for applying the name Bellatrix is obscure. It's a Latin word meaning "Female Warrior".

LORE:

This star's Arabian title meant "The Leader", apt because it's the first of Orion's brightest to rise.

VISUAL ASPECTS:

Bellatrix has an apparent magnitude of 1.6, shining with the blue-white color characteristic of a high stellar range 22,000 K surface temperature.

When observing Orion under dark skies, a curving line of stars traditionally denoting the Hunter's shield may be seen northwest of Bellatrix.

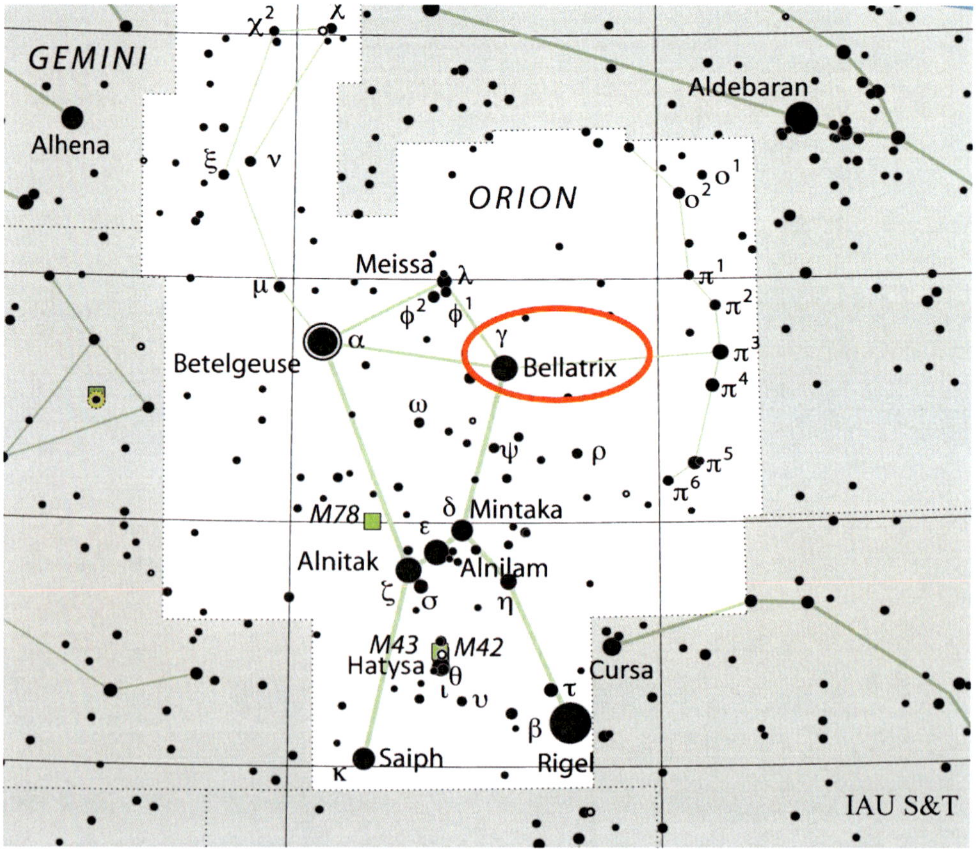

PHYSICAL NATURE:

Distance of Bellatrix from Earth is estimated at about 250 light years and its luminosity is 1,200 times that of the Sun.

Comprising mass equal to 8.4 Suns this B2 spectral type, luminosity class V main sequence star, has six times the diameter of the Sun. That girth represents a volume 216 times larger than the solar capacity.

Bellatrix appears to be surrounded by an expanding shell of gas indicating ejection of photosphere materials into space.

BELLATRIX - ADDITIONAL DATA:

Catalogue Designations: Amazon Star, Gamma Orionis, γ Ori, 24 Orionis, HR 1790, HD 35468, SAO 112740, HIP 25336

****

Parallax: 13 mas
Distance: 77 parsecs, 250 light years
****
Apparent magnitude: 1.6
Absolute magnitude: -2.84
Luminosity: 1,200 solar
Luminosity class: main sequence V
****

Apparent color blend: blue-white
Peak color: blue-white
Color index (B-V): -0.22
Spectral type: B2
Surface temperature: 22,000 K
****
Diameter: 6 solar
Mass: 8.4 solar

****
Coordinates: (J2000.0) 05h25m +06°21' / 81.28 +6.35
Transit at 9 p.m. January 27
Proper motion: -8 -13 mas/year
Radial velocity: +18 km/sec.
****

The B2 spectral type designation of Bellatrix indicates that when we see this star on a clear night it will appear blue-white, with this color corresponding to a surface temperature of about 22,000 Kelvins – far hotter than the Sun's surface – which is about 5,800 Kelvins. The OBAFGKM array of star color designations developed at the Harvard College Observatory in the late 19th and early 20th centuries under the directorship of Edward C. Pickering. During their *Henry Draper Memorial* project hundreds of thousands of stars had their individual spectra photographed wholesale mainly using two similar telescopes, one at Harvard and the other in Peru with 8-inch diameter objective lenses and 45-inch focal lengths. The objective lenses were covered by a large 13-degree angle prism, which thereby produced small spectra of each star in the photograph's entire field of view rather than the multitude of pin-point spots seen in "normal" wide-angle star photographs. Thereby, the spectra of hundreds of stars might be seen on just a single photograph.

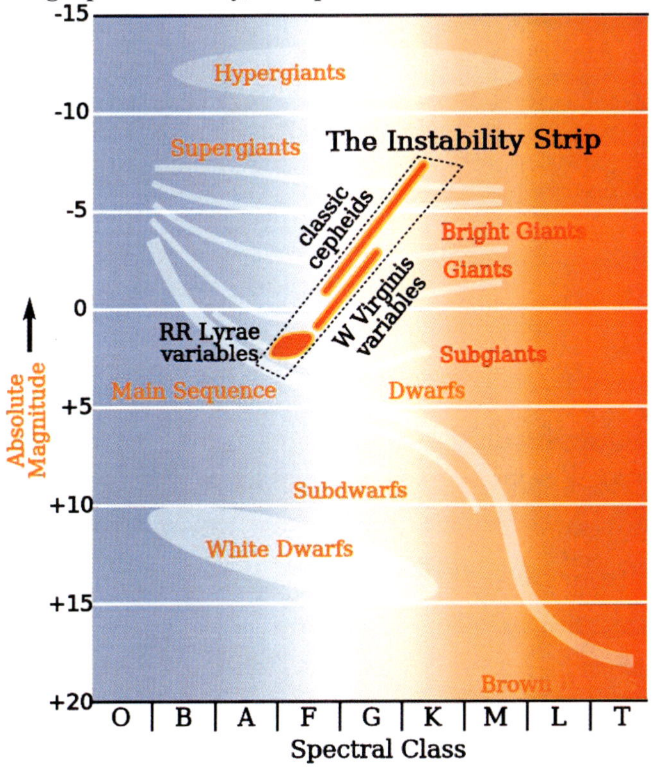

Example of an H-R diagram Credit: Rursus

After the glass photographic plates mostly 8 x 10 inches in size were developed, they moved to the bailiwick of "Women Computers" a highly skilled team that did a Herculean job over many decades discerning and categorizing hundreds of thousands of obscure little smudges of black and white detail. After many iterations based on emerging insights, a system crystalized, categorizing stars based on their colors/surface temperatures, ranging from the hottest on the left to "coolest" on the right. The array was divided into compartments given designations: O, B, A, F, G, K, and M, with each "bin" divided from left to right into 10 compartment sub-divisions. Therefore, we may see that with spectral type B2, Bellatrix is about 20% of the "way" between spectral type B0 and type A0.

In the second decade of the Twentieth Century, Danish astronomer Ejnar Hertzsprung and American Henry Norris Russell independently gleaned information about the intrinsic brightness of a host of stars based on rather rough correlations between the stars' apparent brightness seen from Earth and the so-called "proper motions of those selected stars. The concept both astronomers relied on was the assumption that the further a star is from Earth the smaller will be its proper motion across the sky as observed over many decades by generations of astronomers. Stars with relatively large proper motions had their brightness "handicapped" and stars with smaller proper motions were given increases to thereby more equitably give an estimate of their "real" relative brightness, un sullied by the unfortunate fact that some appeared to have a brightness advantage because they are closer to Earth! The "adjusted" "fair" estimate of a star's brightness came to be known as its absolute magnitude, in contrast to its apparent magnitude – how bright the star, regardless of distance, appears in the night skies of Earth.

The genius of Hertzsprung and Russell was to independently produce charts showing the absolute magnitude of a selection of stars on the vertical Y-axis and the apparent color of each of those stars assigned to a position on the X-horizontal axis. Due to the dearth of trans-Atlantic professional communications during World War I, they were unaware of each other's work in this field.

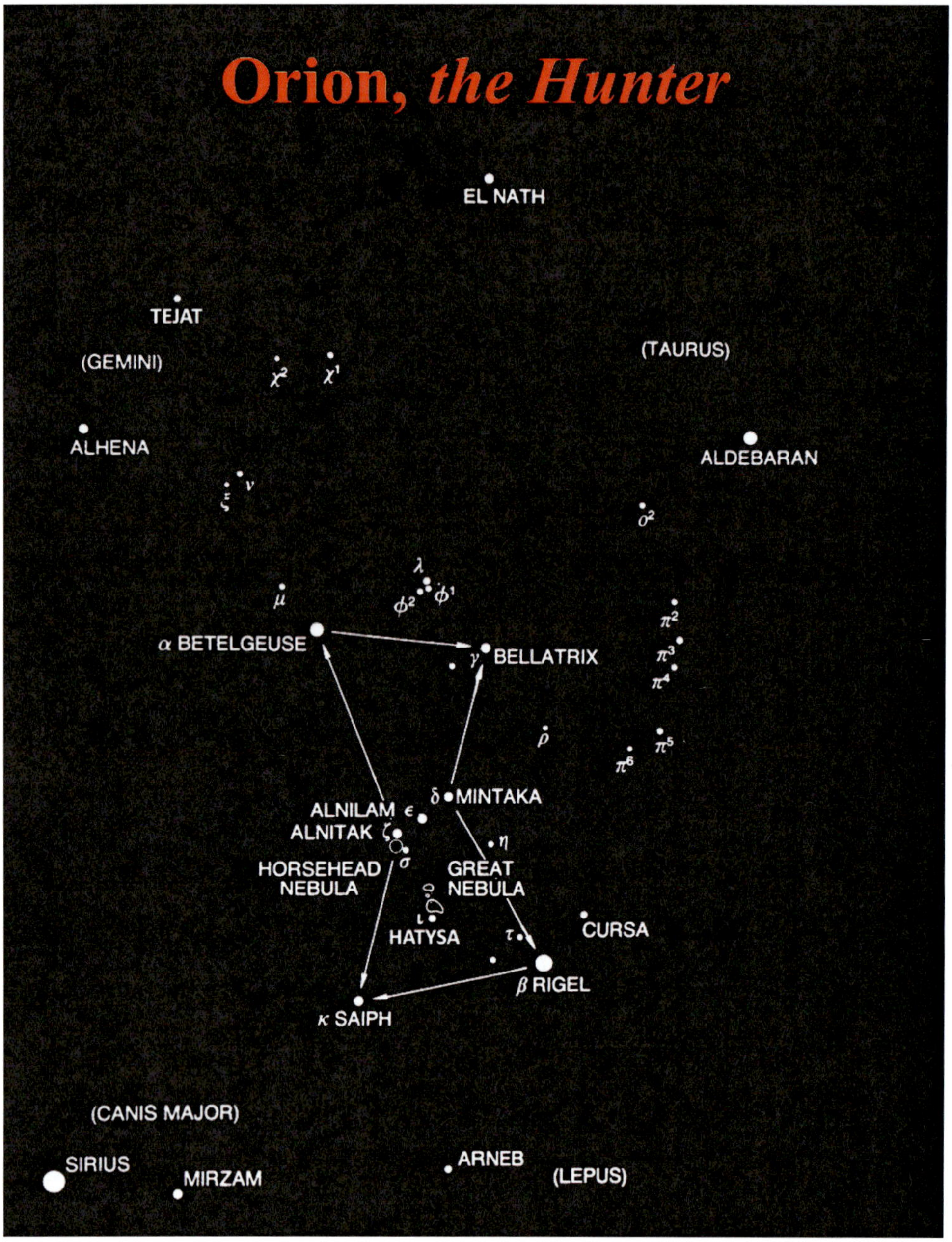

oh-RYE-un, uh-RYE-un
Ori – Orionis, ***or-eye-OH-niss***
>>>>>>>>>>>>>>>>>>>>>>>
Brightest Stars
December 1 Bellatrix 1.6 blue-white main sequence 250 ly
December 8 Betelgeuse 0.42 yellow-orange supergiant 490 ly
December 9 Mintaka 2.4 blue-white/blue-white giant/main sequence 690 ly
December 10 Alnilam 1.7 blue-white supergiant 2,000 ly
December 13 Alnitak 1.9 blue-white supergiant 740 ly
December 14 Rigel 0.1 blue-white supergiant 860 ly
December 16 Hatysa 2.8 blue-white/blue-white giant/giant 2,300 ly
December 24 Saiph 2.1 blue-white supergiant 650 ly
>>>>>>>>>>>>>>>>>>>>>>>
EVENING SEASON:

December - March

SKY TRACK:

Orion is an equatorial constellation bisected by the celestial equator. It rises around the eastern point of the horizon, crosses the meridian about halfway between our horizon and zenith, and then sets directly to the west.

LORE:

Derivation of the name Orion is veiled in antiquity. It may have come in part from association with a Mesopotamian god Urn-ann a, known as the "Light of Heaven". Ancient Israelites saw these stars as the "Giant", "Nimrod"; and until about 500 B.C., Orion was known in Greece simply as the "Warrior". By the time of Athenian writer Euripides (480-406 B.C.) the name Orion had come into general use.

Predawn (heliacal) rise of Orion now coincides with the peak of summer heat. The constellation's midnight rise takes place when

forests in the North Temperate Zone show their glorious color changes, and in December Orion makes its evening debut at the start of the frosty holiday season.

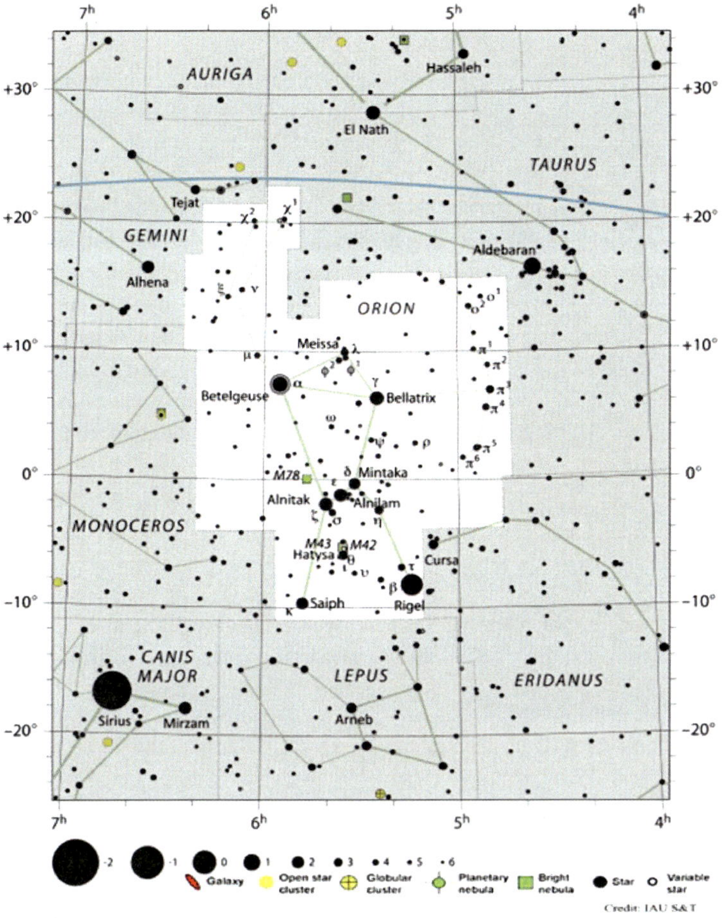

Greek mythology describes Orion as a prince of Boeotia, an ancient kingdom northwest of Athens. Orion's birth was a reward to his father, King Hyrieus, for his hospitality to gods Zeus, Hermes, and Poseidon, who had appeared incognito as poor strangers in his kingdom. Legends tell us that Orion grew to such great size that he could walk across the sea floor without getting his ears wet. He became very strong, a mighty hunter who was followed everywhere by his faithful dogs.

After many adventures, Orion traveled to Crete where he befriended Diana, goddess of the Moon and of the hunt. They fell in love and spent much time together on Crete. Therefore, Diana neglected her responsibility guiding the lunar chariot across the night sky. Her brother Apollo, driver of the solar chariot, warned Diana to return to her duties, but she would not leave Orion. In time, Apollo tricked her into killing Orion with an arrow as he swam far from shore. After this tragedy, Diana transported Orion to the sky as the most glorious of constellations.

VISUAL ASPECTS:

Brilliant stars of Orion are highlights of a section of the local Milky Way known as the Orion Arm images. This is the only constellation which presents two first-magnitude stars, Betelgeuse and Rigel. They provide the best example of stellar color contrast visible to unaided eyes. Betelgeuse is called a red giant, but actually appears approximately yellow orange. Rigel diagonally across the constellation figure from Betelgeuse is blue-white.

Approaching the Holiday Season, stars in and around magnificent Orion begin dominating our views of the evening sky. Some features are famous, with names plucked from astronomical charts and texts and then applied in other realms of endeavor, such as Sirius, Betelgeuse, Subaru, and Orion itself.

Orion Stars and Nebulosity, punctuated by red light from hot hydrogen emission nebulae – Credit: Rogelio Bernal Andreo

NEARBY FEATURES: **A binocular reprise of this constellation's delightful vista is a wonderful warm-up to a winter evening's celestial encounters. Check out Steve's Hayden Planetarium Blog from Jan 11, 2016 regarding** ***Orion and Company at Center Stage.***

Tejat December 3

East-Northeast East East-Southeast

Gemini

December – May

TAY-jat

2.9v mag. yellow-orange giant; 230 ly; pulsating variable star

****

Visualizing the Sun as the size of a baseball, by comparison Tejat would be about 12 feet wide – two-thirds the width of a pitcher's mound. A scale model of Tejat then would be about 71,650 miles from the baseball/Sun. That's approximately equal to a great circle trip three times around the world, or 30% of the distance from Earth to the Moon.

Alternately, if the Sun had the diameter of a soccer ball, another Tejat model, at this scale, would be 11 meters wide – 60% the size of a soccer field's center circle. Scaled distance from our soccer ball/Sun to this Tejat model would be 343,850 kilometers, 89% of the average distance between Earth and the Moon.

LOCATION:

Third magnitude Tejat is 15 degrees southeast of El Nath.

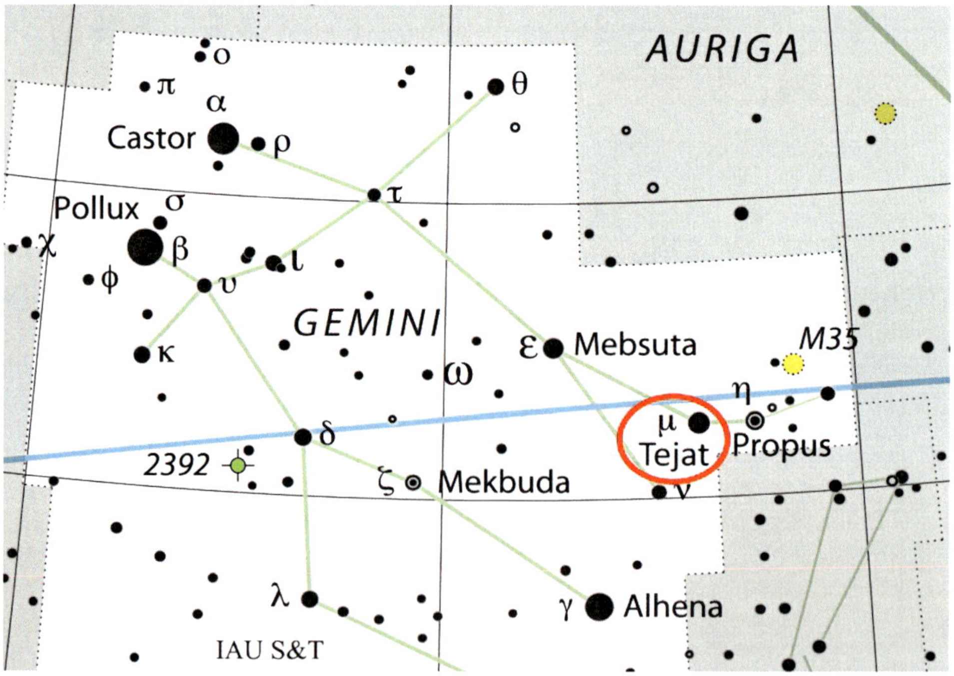

LORE:

This star was described in Tycho Brahe's star catalogue as the "Heel of Castor", in the western "Twin of Gemini". The name Tejat Posterior had been used in the past for this star, indicating the "Back Foot".

Tejat was approved as the official name for Mu Geminorum by the International Astronomical Union (IAU) in 2016.

VISUAL ASPECTS:

Tejat has average apparent magnitude 2.9 and appears to the eye with a yellow-orange color. Its distance from the Earth is about 230 light years, and this star's luminosity is about 300 times greater than our Sun's intrinsic brightness. As is the case with many giant stars of this spectral type, Tejat is an irregular variable. Its magnitude range is between 2.75 and 3.02.

PHYSICAL NATURE:

This spectral type M3 luminosity class III giant star has a mass 2.1 times that of the Sun. The diameter of Tejat is 50 solar, its volume is 125,000 times the Sun's size, and the B-V color index is 1.61. Estimated surface temperature of Tejat is 3,800 K, near the low end of the stellar range.

TEJAT - ADDITIONAL DATA:

Catalogue Designations: Mu Geminorum Aa, μ Gem Aa, 13 Geminorum, HD 44478, HIP 30343, HR 2286, SAO 78297

Parallax: 14 mas
Distance: 71 parsecs, 230 light years
****
Apparent magnitude: 2.9v (2.75-3.02)
Absolute magnitude: -1.4
Luminosity: 300 solar
Luminosity class: giant IIIv
****

Apparent color blend: yellow-orange
Peak color: red
Color index (B-V): 1.61
Spectral type: M3
Surface temperature: 3,800 K
****
Diameter: 50 solar
Mass: 2.1 solar

****
Coordinates: (J2000.0) 06h23m +22°31' / 95.74 +22.51
Transit at 9 p.m. February 11
Proper motion: +56 -110 mas/year
Radial velocity: +54 km/sec.
****

NEARBY FEATURES:

The Summer (Northern) Solstice (6h00m +23°24') – The Sun reaches its most northerly point on the ecliptic about five degrees west of Tejat. This location known as the Summer Solstice is defined on the celestial sphere by intersection of the six-hour coordinate of right ascension with the line of the ecliptic. Currently, the center of the Sun's disk, as summer begins in the Northern Hemisphere and winter in the southern, is at the border between Gemini and Taurus, as defined by the IAU constellation boundaries of 1930.

M35 open star cluster Gemini (6h09m +24°20') – This attractive cluster, designated number 35 on Charles Messier's list has a position 3.5 degrees northwest of Tejat and two degrees northeast of the ecliptic's summer solstice point. M35 is visible through binoculars and resolvable to individual stars by observers with small telescopes.

Messier 35 contains nearly 500 stars, distributed over a volume of space about 30 light years in diameter. At an estimated distance of 2,800 light years from the Solar System, M35 covers a patch of sky equal to the Moon's disk. A low-power eyepiece is needed to fit M35 in a telescopic field of view.

Eagle-eyed observer Steven James O'Meara (1956-) notes aspects of M35's appearance in his excellent book, *Deep-Sky Companions: The Messier Objects.* He recommends viewing it and other open clusters through a rich-field telescope to best appreciate their appeal.

Hubble's Variable Nebula NGC 2261 C 46 Monoceros (6h39m +8°45') About 15 degrees south of Tejat this strange triangular nebulosity extends northward like a fan, from the variable star R Monocerotis. It was discovered by William Herschel in 1783.

Edwin Hubble discovered that NGC 2261 varies in brightness and structure as a result of photographic comparisons he made during 1916. Observations from recent decades suggest the central star associated with the nebula is either a T Tauri type nascent variable or perhaps a protostar at an even earlier stage of gravitational contraction. Distance of this nebula to the Solar System is approximately 2,500 light years.

The Rosette Nebula Monoceros (6h34m +5°00') – A doughnut shaped region of glowing gas, covering an area of sky about three times the size of the full Moon's disk, the Rosette is 20 degrees south of Tejat. Its light is faint, barely visible with binoculars even under the best conditions. Long-exposure photographs reveal the Rosette as a beautiful complex structure. It's considered the prime example of a Stromgren sphere, the circumstellar zones of ionized hydrogen described in 1939 by Danish astronomer Bengt Stromgren(1908-1987).

At a distance estimated to be 5,200 light years from us, the Rosette Nebula's diameter is about 120 light years. It encompasses young open star cluster NGC 2244, centered in the Rosette. Formation of NGC 2244's stars consumed much dust and gas in the nebula's central zone. Stellar winds from those young stars then blew away most of the remaining nebulosity, leaving the center of the Rosette relatively clear of such interstellar matter.

The Rosette Nebula and open star cluster NGC2244 – asi

Gemini, the Twins

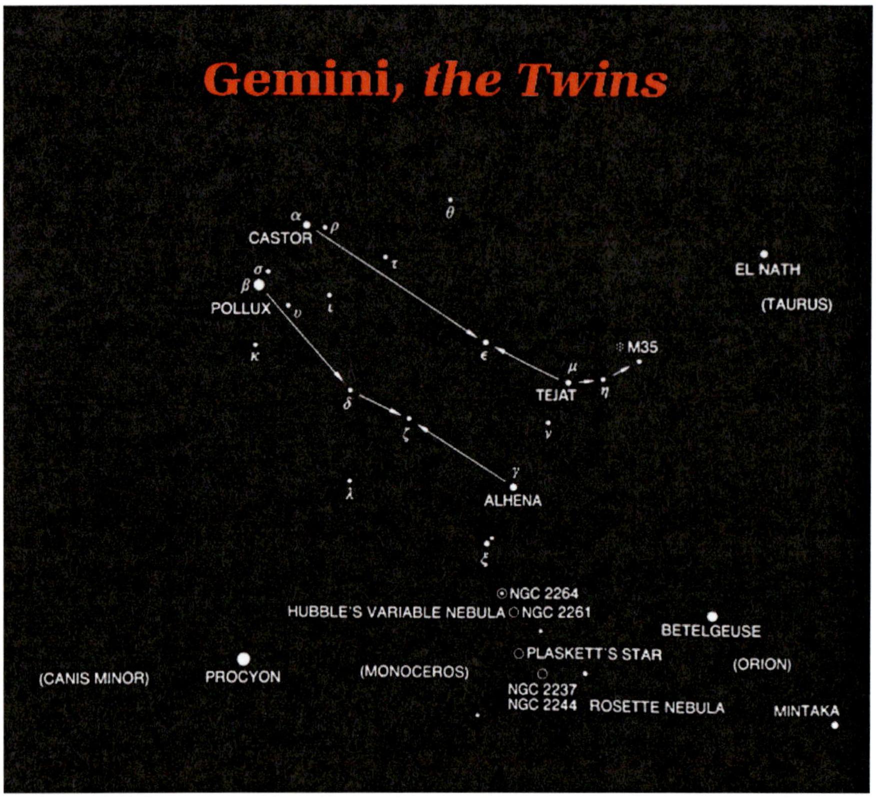

JEM-uh-nye, JEM-uh-nee
Gem – Geminorum, ***JEM-uh-NOR-um***
>>>>>>>>>>>>>>>>>>>>>>

Brightest Stars:

December 3 Tejat 2.9 yellow-orange giant 230 ly

December 12 Alhena 1.9 blue-white subgiant 110 ly

December 17 Castor 1.6 blue-white main sequence 51 ly

December 21 Pollux 1.1 yellow giant 34 ly

>>>>>>>>>>>>>>>>>>>>>>

EVENING SEASON:

December - May

SKY TRACK:

Gemini rises in the northeast, transits the meridian high in the southern sky, and sets toward northwest.

LORE:

Castor and Pollux, prime features of this constellation, form a striking pair known as "Twin Stars" in many lands. Castor and Pollux are Greek names for these sons of Leda, Queen of Sparta. Zeus, under the guise of a swan, came to her and thereby fathered immortal Pollux, the boxer. King Tyndareus of Sparta was father of mortal half-brother Castor, the horseman.

The Gemini brothers were sometimes known as the "Dioscuri", with both regarded as sons of Zeus. Castor and Pollux were members of the Argonautic expedition seeking the Golden Fleece. The Twins of Gemini were considered patrons of navigators and warriors. As such, their images were carved into figureheads for ships; and they served as symbolic heralds for Roman armies. A major structure in the Forum, the Temple of Castor and Pollux, is today a striking ruin, its columns symbols of vitality and hope after so many ages. In Israel, the Twins were said to represent Simeon and Levi, the "Brethren".

VISUAL ASPECTS:

Gemini, northeast of Orion with its stars at the eastern fringe of the Milky Way, is the Sun's location during first weeks of summer. Stellar lines extending westward from both Castor and Pollux can be imagined as representing bodies of the Twins.

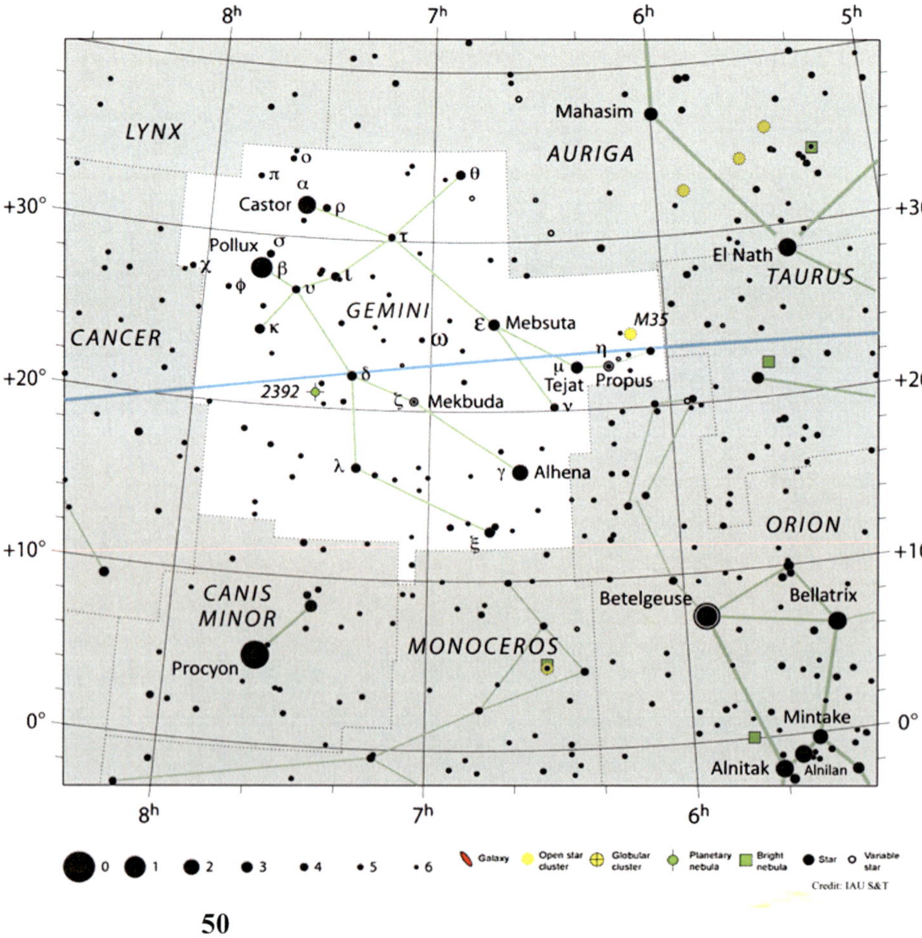

Cursa December 7

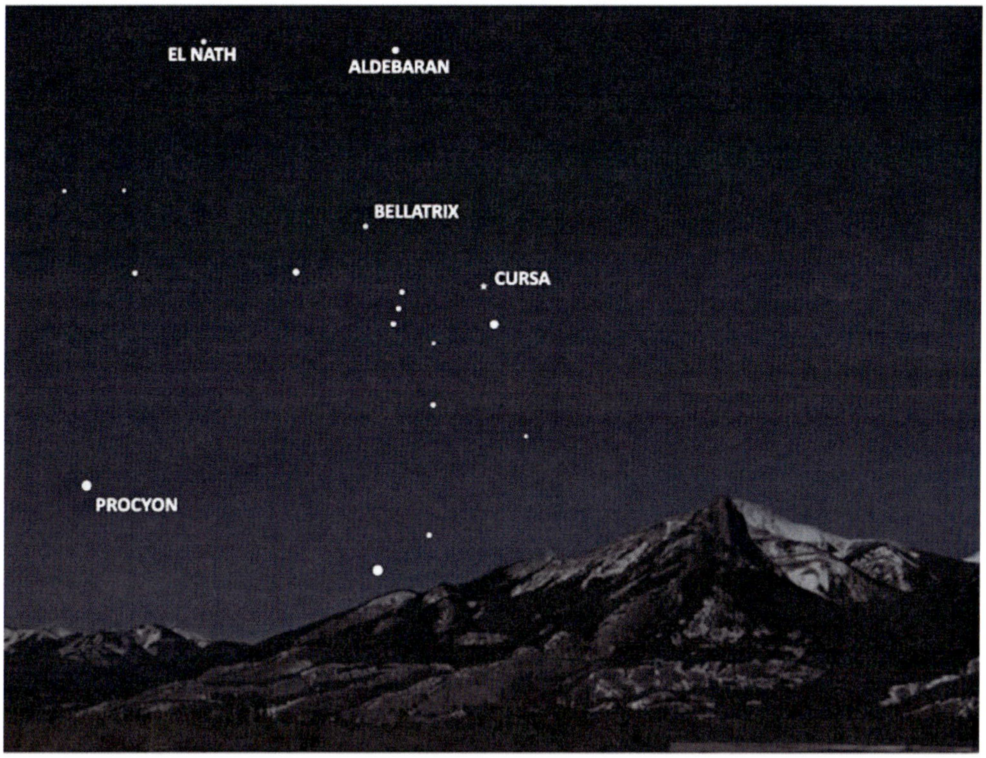

East-Southeast Southeast South-Southeast

Eridanus

December – March

KUR-sah

2.8 mag. blue-white subgiant; 89 ly

****

If we imagine the Sun being the size of a baseball, on that scale Cursa would be about seven inches wide. Our model of this star would have to be about 27,700 miles from the baseball/Sun, that's just a bit further than a trip around the world.

Alternately, with the Sun having the diameter of a Soccer ball, a Cursa model at this scale, would be 53 centimeters wide. Corresponding distance from the soccer ball/Sun to our metric Cursa model would be about 133,100 kilometers, equivalent to three and a third trips around the Earth, and 35% of the actual average distance from Earth to the Moon.

LOCATION:

Third-magnitude Cursa is three and a half degrees of arc northwest of brilliant Rigel, the brightest star in Orion.

MEANING:

Cursa, along with neighbors Lambda, Psi, and Omega Eridani were known in Arabia as the "Footstool of Orion" and its name is adapted from that title.

LORE:

This star marks the eastern end of long winding constellation Eridanus, known as the River. Before influences of Greek astronomy became prevalent, Cursa and its two neighbors were earlier known in Arabia as the "Ostrich Nest". It is said that in ancient China Cursa was called the "Golden Well".

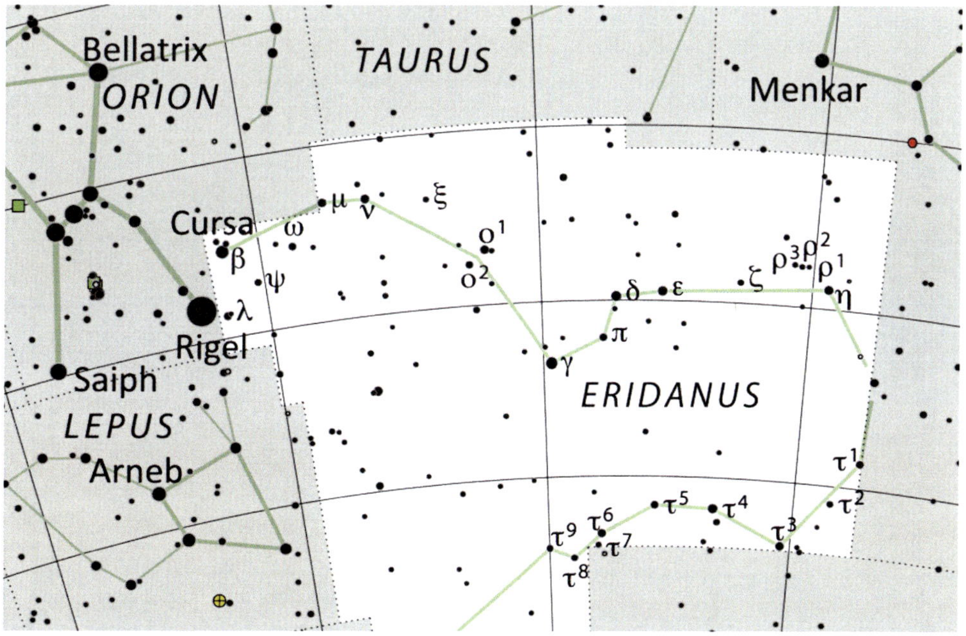

VISUAL ASPECTS:

Cursa shines with apparent magnitude 2.8 as a blue-white subgiant. Its spectrum indicates a surface temperature of about 8,400 K.

PHYSICAL NATURE:

Subgiant Cursa has spectral type A3, and a blue minus visual (B-V) color index of 0.13. It comprises mass twice the solar amount, and it's 2.4 times the Sun's diameter. Cursa's volume therefore is 14 times larger than the Sun.

The trigonometric parallax of Cursa is 37 milliarcseconds, indicating a distance of 27 parsecs, about 89 light years. With that separation from Earth, the star's apparent magnitude computes to luminosity about 48 times more intense than the Sun.

CURSA - ADDITIONAL DATA:

Catalogue Designations: Beta Eridani, β Eri, 67 Eridani, HD 33111, HIP 23875, HR 1666, SAO 131794

****

Parallax: 37 mas
Distance: 27 parsecs, 89 light years

****

Apparent magnitude: 2.8
Absolute magnitude: 0.6
Luminosity: 48 solar
Luminosity class: subgiant IV

****

Apparent color blend: blue-white
Peak color: white
Color index (B-V): 0.13
Spectral type: A3
Surface temperature: 8,400 K

****

Diameter: 2.4 solar
Mass: 2.0 solar

****

Coordinates: (J2000.0) 05h08m -05°05' / 76.96 -5.09
Transit at 9 p.m. January 23
Proper motion: -83 -75 mas/year
Radial velocity: -6 km/sec.

****

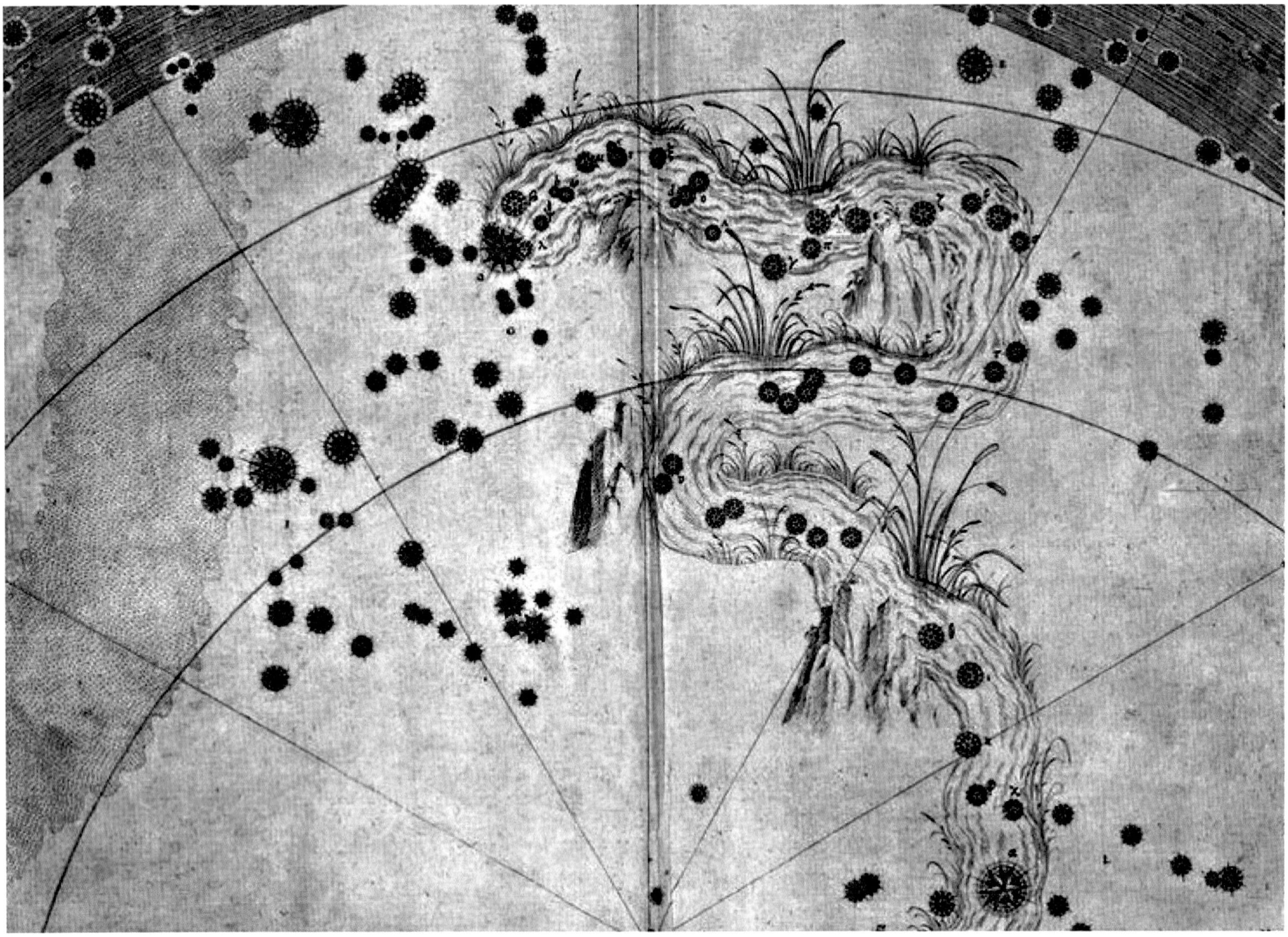

Eridanus – Bayer, Uranometria 1603

Eridanus, *the River*

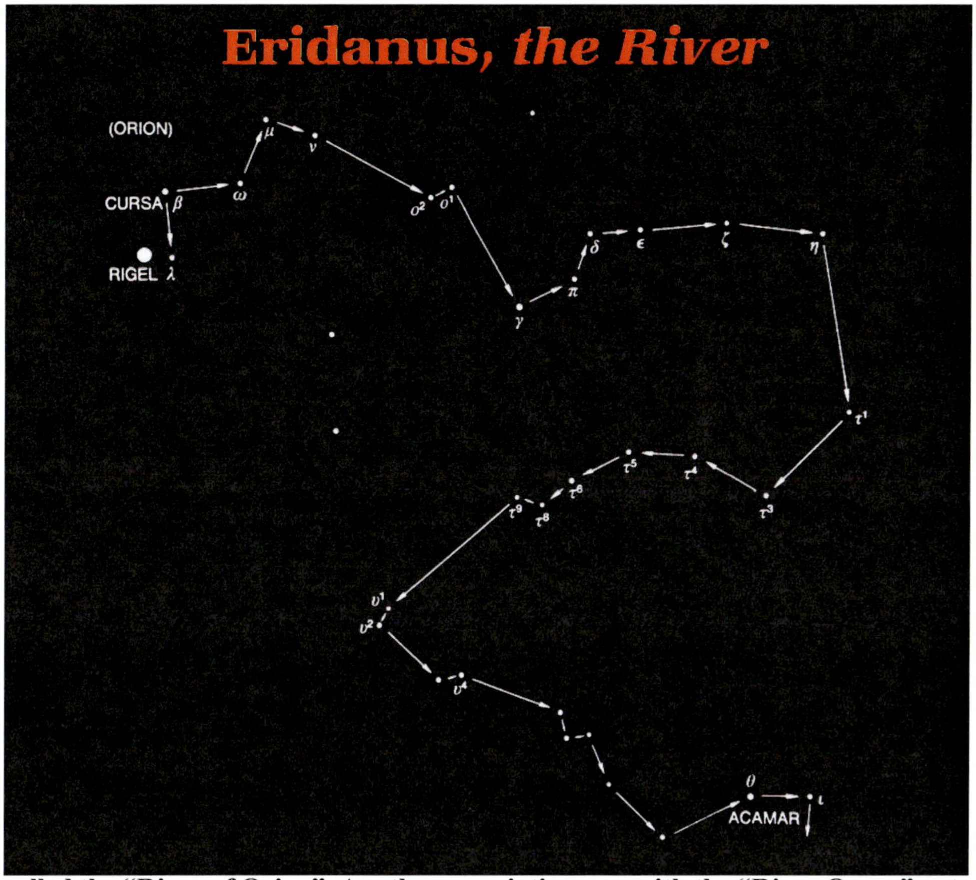

ih-RID-un-us
Eri – Eridani, *ih-RID-un-eye*
>>>>>>>>>>>>>>>>>>>>>
This Constellation's Brightest northern star (visible from mid-northern latitudes):
December 7 Cursa 2.8 blue-white giant 89 ly
>>>>>>>>>>>>>>>>>>>>>

EVENING SEASON:

December - March

SKY TRACK:

Eridanus follows a southerly sky track after rising along the southeastern horizon before setting in the southwest.

LORE:

The chain of stars forming Eridanus was known as a river throughout the ancient Mediterranean and Middle Eastern worlds. It was frequently associated with a large river in a particular region such as the Nile, Euphrates, or Po. Since one end of Eridanus is near Rigel, it has been called the "River of Orion". Another association was with the "River Ocean" noted by Homer and believed to flow around the entire ancient world.

VISUAL ASPECTS:

From third-magnitude Cursa near Orion, stars of Eridanus wend westward, then meander south and east, finally disappearing below the horizon for observers at mid-northern latitudes.

Eridanus is the longest, albeit serpentine, constellation and one of the faintest. However, its southern terminus, compensating a bit by its brilliance, is vivid Achernar, with magnitude 0.5 at declination -57 degrees. That's nearly as far south as the Southern Cross, and neither sky feature clears the horizon for observers at mid-northern latitudes.

The Front Gallery image on page ix shows an area of extremely distant young galaxies located in the direction of the constellation Fornax.

The image's name is ASPIRE, J0305-3150, indicating its location in the constellation Fornax at Right ascension 03h 05m, declination 31° 50'. The "X" in *FORNAX* seen in the map to the right marks the approximate site of this ASPIRE image.

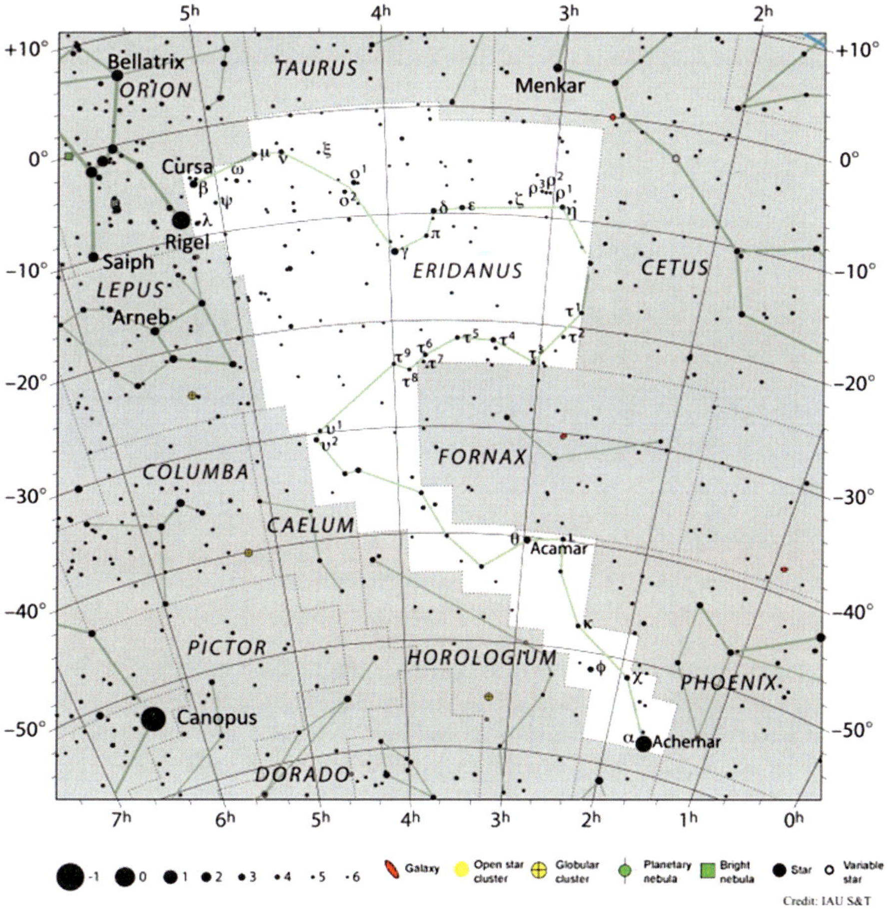

Betelgeuse December 8

East East-Southeast Southeast

Orion

December – March

BEETLE-juice

0.42v mag. yellow-orange supergiant; 490 ly; irregular long period pulsating variable star

Anchor Point Star – M1-M2 Ia-Ib

****

Imagine the Sun being the size of a baseball. With that scale Betelgeus*e* would, at its average size during pulsations, be approximately 260 feet wide – 14 times the width of a pitcher's mound, twice the diameter of the infield diamond, and about 65% the size of a typical Major League playing field.

At that scale, our Betelgeuse model would have to be 152,600 miles from the baseball/Sun model, a span equal to a journey six times around the Earth, or 64% of the Moon's average distance from us.

On the other hand, if soccer is your game, with our imagined Sun having the diameter of a soccer ball, the corresponding average size Betelgeus*e* model would be 237 meters wide – 13 times the size of a soccer field's center circle, over twice the length of a standard pitch, and as wide as the entirety of any world class football stadium.

Scaled separation of our huge Betelgeuse model from a soccer ball sized Sun would be about 732,600 kilometers, equivalent to 18 flights around the world, and nearly double the average distance between Earth and Moon.

Quite a star!

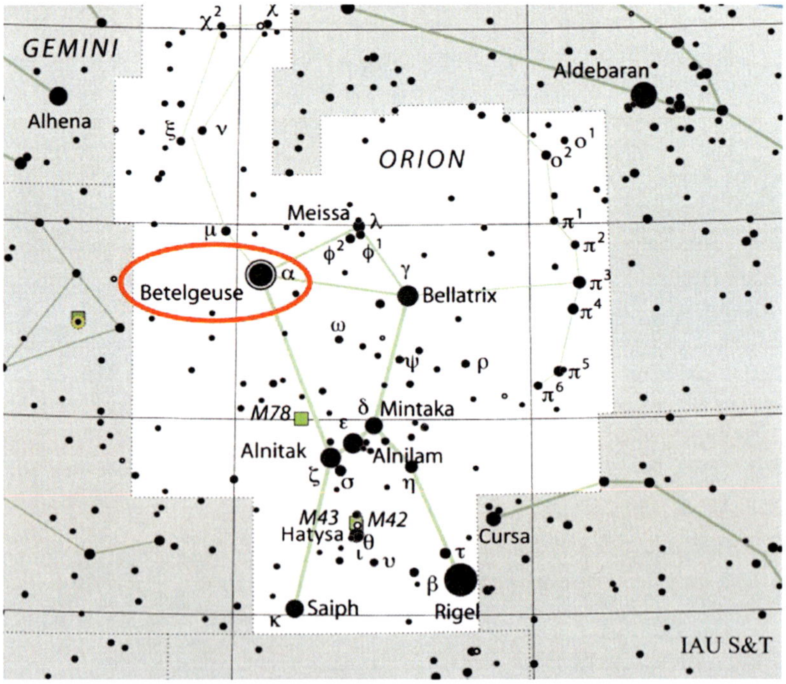

LOCATION:

This marvelous first magnitude supergiant shines at the shoulder of Orion, seven degrees east of Bellatrix.

MEANING:

The name Betelgeuse is very loosely derived from an early Arabian description of this star's position, marking the "Hand of the Central One", referring to its location in the striking figure of Orion. Paul Kunitzsch (1930 – 2020), a scholar at the University of Munich who did extensive studies of star name etymologies, found that when the star's Arabic title was transliterated in Paris during 1246 the translator incorrectly used a B instead of a Y at the beginning of its name – an error that was perpetuated. In later centuries researchers had searched in vain for origins of "Betelgeuse", essentially an artificial appellation.

Kunitzsch believed that around the year 1600 Joseph Scaliger (1540-1609) introduced the current spelling of Betelgeuse; reflecting what he incorrectly thought was the star's original Arabic designation, meaning "Armpit of Orion".

LORE:

Betelgeuse and some of its stellar neighbors were known in Mesopotamia as the "Constellation of the King". In India, Betelgeuse was called the "Arm", a title also used by Persians. In ancient Greece, Betelgeuse marked the "Right Shoulder

of Orion", the star's designation in Ptolemy's *Almagest* (*Syntaxis*) of the second century A.D. Betelgeuse and its neighbors form for us the most captivating array of stars in the entire sky. Their brilliance was noted in 424 B.C., by words of Thracian king Polymester, in Euripides (c.480-c.406 B.C.) *Hecuba*:

"Whither can I turn or go? Shall I take wings and soar aloft to the mansions of the sky, where Orion and Sirius dart from their eyes a flash as of fire"

HISTORY:

In 1835 science philosopher Auguste Comte (1798-1857) writing about stars in *Cours de Philosophie Positive* conceded: *"...the possibility of determining their shapes, distances, sizes, and their movements".* However, he went further and firmly asserted:

"We would never know how to study by any means their chemical composition, or their mineralogical structure, and even more so, the nature of any organized beings that might live on their surface."

John Hearnshaw, University of Canterbury, Christchurch New Zealand, notes "Comte was in his time an influential philosopher of science, but if he's remembered today it's mainly because of those remarks regarding stellar composition." Comte did not foresee implications arising from the remarkable work of his contemporary Joseph Fraunhofer (1787-1826).

Prof. Hearnshaw, an expert in stellar spectroscopy, reminds us that decades before Compte wrote his inopportune quotable quote, Fraunhofer had already discovered a host of parallel dark lines spread across rainbows of color spectra from the Sun, Betelgeuse, and Sirius. Decades later those lines became essential tools for chemical and physical analysis of stars.

It would take others, beginning with Robert Bunsen (1811-1899) and Gustav Kirchhoff (1824-1887) to begin unraveling starlight chemical mysteries. Austrian Christian Doppler (1803-1853) developed means of using frequencies of starlight to disclose information about stellar motion (also providing us with knowledge about speeds of tornado spins and fastballs). Full application of Fraunhofer lines to understand stellar fundamentals finally became possible after 20th century discoveries in quantum physics.

Betelgeuse was first described as a variable star by John Herschel in 1836. In 1849 he wrote that the star's changes in magnitude were especially noticeable from 1836 until 1840. After variability settled down for a while it again increased, and by the end of December 1852 Herschel considered Betelgeuse the brightest star in the night sky north of the celestial equator.

During 1920, an innovative 20 foot wide interferometer installed atop the 100-inch reflector on Mt. Wilson enabled Albert Michelson (1852-1931) and Francis Pease (1881-1938) to directly measure the diameter of Betelgeuse. They reported its angular girth at 47 milliarcseconds (mas), with an uncertainty range of 10%.

Introduction of speckle interferometry in the 1970's revealed an image of Betelgeuse's shape – the first ever "close-up" portrait of a distant star. Then in 1995 the Hubble Space Telescope directly imaged Betelgeuse, providing the first actual picture obtained of a stellar disk other than that of the Sun.

The European Southern Observatory's Very Large Telescope in 2009, equipped with the NACO adaptive optics instrument and using lucky imaging techniques, produced a near-infrared direct image of Betelgeuse.

During 2011, Pierre Kervella of the Paris Observatory and his team, reported in the Journal Astronomy and Astrophysics on various structures seen across Betelgeuse's photosphere. Additionally, they described the enormous asymmetric gas envelope surrounding this star. That circumstellar region features a bright gaseous plume extending six times the star's radius from the photosphere, the star's visible surface.

VISUAL ASPECTS:

Betelgeuse varies in brightness over the course of a semiregular period and on average is the seventh-brightest star in our night sky. Appearance ranges from about maximum magnitude +0.3 to +1.2 minimum magnitude over periods of approximately 6.4 years. That represents an apparent brightness intensity change of 275 percent.

This Alpha star of Orion is one of many whose brightness is regularly monitored by American Association of Variable Star Observers. Luminosity of Betelgeuse varies as the star expands and contracts by about ten percent of its average diameter. Although red is the most intense individual color in its visible spectrum, the blend of all colors in the wavelength range appears yellow-orange to most eyes. About 90% of all photons emitted at the surface of Betelgeuse have invisible infrared wavelengths.

During 2019 and 2020 Betelgeuse dramatically faded to magnitude 1.7+, but by early 2021 rebounded to about magnitude +0.6. The cause of this significant variation remains uncertain, and it has variously been attributed to substantial ejections of obscuring dust, and unusual star spot activity.

PHYSICAL NATURE:

Betelgeuse is a vast supergiant. During its size variations, its diameter ranges between about 950 and 1,200 times that of the Sun. If it was in the Solar System, at the Sun's location, Betelgeuse at its maximum radius would extend past Jupiter's orbit.

A wide zone of expanding shells surrounds Betelgeuse, consisting of various stellar layers previously ejected during episodes of violent activity. Image

Betelgeuse is another Anchor Point star, its spectrum serving as a standard for its type and class since inclusion in the 1943 Morgan, Kennan, Kellman Atlas. It defines spectral type is M2 and luminosity classes Ia-Ib. With its average surface temperature about 3,400 K, nearly 90 percent of this star's energy is radiated as invisible infrared radiation. This vast star is estimated to have a mass between 15 and 20 times that of the Sun. Its estimated maximum diameter indicates a volume about 1.7 billion times larger than the Sun!

BETELGEUSE - ADDITIONAL DATA:

Catalogue Designations: Alpha Orionis, α Ori, 58 Orionis, HR 2061, HD 39801, HIP 27989, SAO 113271

****

Apparent magnitude: 0.42v (0.3-1.20)
Absolute magnitude: -5.5
Luminosity: 13,000 solar
Luminosity class: supergiant Ia-Iab

****

Parallax: 6.6 mas
Distance: 152 parsecs, 490 light years

****

Apparent color blend: yellow-orange
Peak color: red
Color index (B-V): 1.85
Spectral type: M2
Surface temperature: 3,150-3,650 K

****

Diameter: 950-1200 solar
Mass: 15-20 solar
Coordinates: (J2000.0): 05h55m +07°24' / 88.79 +7.41
Transit at 9 p.m. February 4
Proper motion: +25 +8 mas/year
Radial velocity: +21 km/sec.

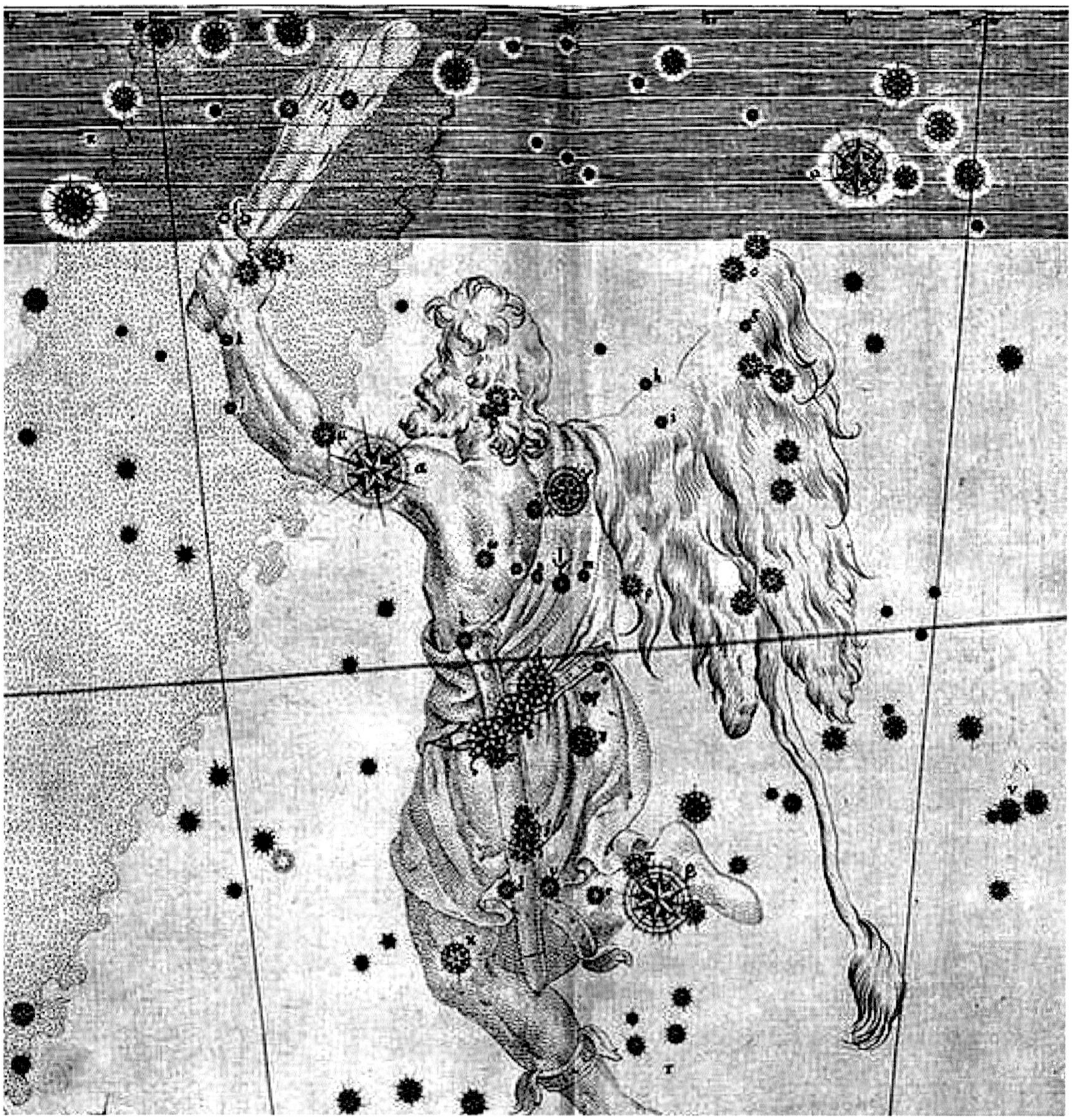

Orion – Bayer, Uranometria, 1603

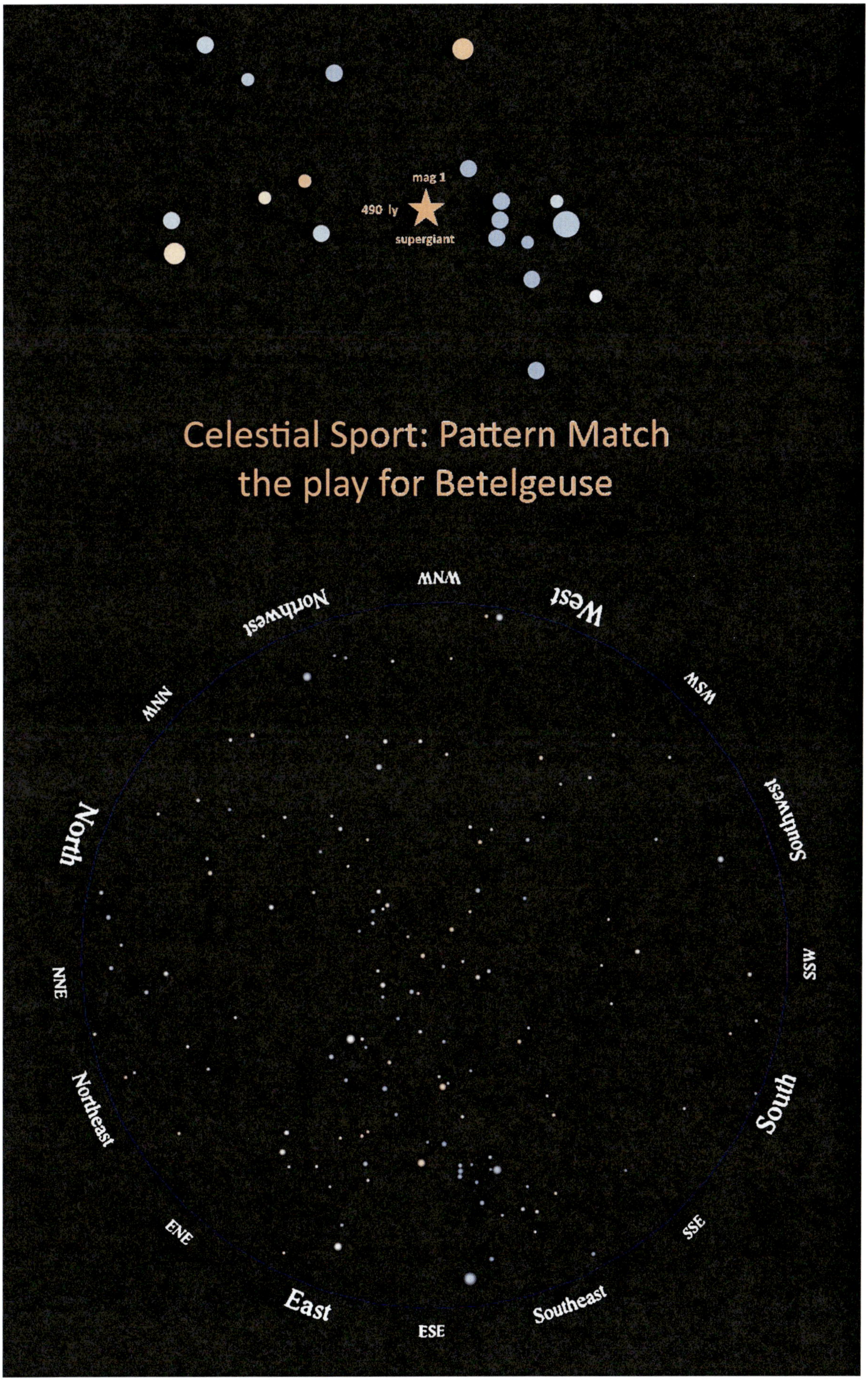
mag 1
490 ly
supergiant
Celestial Sport: Pattern Match
the play for Betelgeuse
WNW
West
Northwest
NNW
WSW
Southwest
North
SSW
NNE
South
Northeast
SSE
ENE
East
ESE
Southeast

Mintaka December 9

East **East-Southeast** **Southeast**

Orion

December – March

min-TAK-ah

2.4 mag. blue-white giant; 690 light years; spectroscopic binary, two eclipsing Algol type components

****

If we envision the Sun as the size of a baseball, at that scale the primary component of the Mintaka multiple star system, a bright giant, would be about four feet wide – one-quarter the width of a pitcher's mound. our model of this star would have to be placed about 392,500 miles from a baseball sized Sun, that's equal to nearly 16 trips around the world, or one and two-thirds times the Moon's distance from Earth.

Considering the Sun as having the diameter of a Soccer ball, another Mintaka model at this scale would be 3.7 meters wide – 20% the size of a soccer pitch's center circle. Scaled distance from the soccer ball/Sun to our Mintaka model would be about 1,884,000 kilometers, equal to almost five times the Moon's actual distance from Earth.

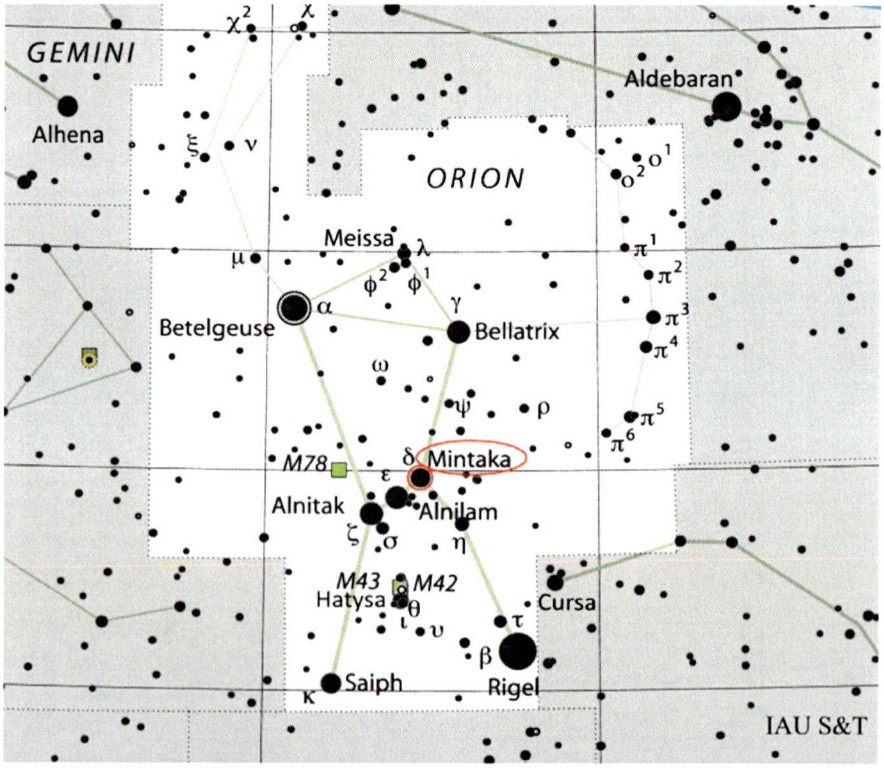

LOCATION:

Mintaka is the westernmost (at the right when the constellation is at the meridian) of three second-magnitude stars forming the "Belt of Orion". It is seven degrees south of Bellatrix.

MEANING:

The name Mintaka means the "Belt", from an Arabian description "Belt of the Central One".

LORE:

Orion's Belt is one of the best-known asterisms. It's formed by a nearly straight line of three stars, almost equal in brightness, separated by 1.5 degrees of arc. Various Arabic name equivalents for these stars include: the "Line", "Golden Grains", "Spangles", "String of Pearls", and "Scale Beam". In China Orion's Belt was called the "Weighing Beam", while on the subcontinent of India, it was known as the "Three-Jointed Arrow".

HISTORY:

The presence of interstellar gas in the Milky Way was discovered from observations of Mintaka's (δ Ori Aa/Ab) spectrum carried out by German astronomer Johannes F. Hartmann (1865-1936) at Potsdam Observatory during 1904. Several components of Mintaka are spectroscopic binary star systems. Spectrum absorption lines of such systems appear to double or at least broaden at regular intervals due to Doppler shifts due to radial velocity differences, as one component

star moves towards us while its companion is moving away. When they regularly cross our line of sight while orbiting their center of mass, no relative Doppler shifts are seen and spectral absorption lines of each component star appear to merge. In the case of Mintaka, Hartmann noted an anomaly in that the calcium K absorption line at 3934 Angstroms wavelength never appeared to split, as did other lines in the Mintaka system's spectrum.

Hartman reasoned the stationary line was produced by an absorption effect from calcium atoms in low density ambient gas clouds distributed along his line of sight to Mintaka. Although calcium absorption is relatively easy to detect, it's a minor constituent of interstellar gas. The mixture of gases found in space between stars of the Milky Way consists mainly of hydrogen. Hartmann's discovery was the first evidence for existence of large quantities of gas in interstellar space.

The first interstellar molecule discovered was methylidyne, CH centered at 4300 Å. It was found by Belgian astronomer Pol Swings (1906-1983) and a colleague physicist L. Rosenthal based on their 1937 studies of spectral profiles observed by means of the spectrograph attached to Coudé focus of the 100-inch telescope at Mt. Wilson Observatory, at that time the world's largest. The spectrum "fingerprint" of this molecule, also known as "Carbyne", was first noted by Fraunhofer in the early 19th century and designated by him as the "G" band.

VISUAL ASPECTS:

Mintaka is part of an array of luminous massive blue-white stars belonging to the Orion OBI Association. Primary pair in the Mintaka star system is a spectroscopic eclipsing binary.

PHYSICAL NATURE:

When we look at Mintaka at the right end of Orion's Belt, we gaze at a complex of five stars in a tight group.

If you aim your telescope at Mintaka, aka Delta Orionis (δ Ori), then look through a 100x eyepiece, three stars appear in a tight group across an area a bit larger than apparent size of Jupiter's disk. The brightest of this set is designated δ Ori A; it contributes almost all the light we see with unaided eyes sparkling at the western end of Orion's Belt.

Our telescopic view also includes a star designated δ Ori C. It consists of a blue-white B2 V main sequence star, in a spectroscopic binary system pairing with a type A main sequence companion. Their orbital period is 30 days and presents a combined visual magnitude of 6.8. This pair is separated from the bright Component A by 53 arcseconds. Gaia archive reports its distance from the Solar System as 1,260 light years.

The faintest component of Mintaka is δ Ori B, reported by Gaia as having visual magnitude 9.2 (although listed in most literature as 14th magnitude). Component B is separated by 33 arcseconds from δ Ori A, Mintaka's second magnitude primary stellar set.

That primary component δ Ori A, in turn comprises three individual stars including an extremely close eclipsing pair designated δ Ori Aa. Those two are orbited by an individual star designated δ Ori Ab, a B0 type subgiant IV, about 1.3 magnitudes fainter than Aa. Separation between the binary δ Ori Aa and δ Ori Ab is 0.267 arcsecond, and their orbital period is listed at 346 years. All three are about 1260 light years from the solar system.

Finally – the two stars contributing most of the eye-catching light when we see at the western end of Orion's Belt are the pair in the δ Ori Aa semi-eclipsing system. These have designations δ Ori Aa1 and δ Ori Aa2. The brighter component is δ Ori Aa1, a type O9.5 II bright giant in orbit with a main sequence star of spectral type B1 designated δ Ori Aa2. These orbit their common center of mass over a period of 5.73 days. The axis of revolution leans at an angle of just 13.7° from vertical, ensuring each star in δ Ori Aa periodically blocks some of its neighbor's light from reaching us. This semi-eclipsing system of Mintaka is similar the well-known Algol binary in Perseus.

The primary star δ Ori Aa1 is an evolved giant of 24 solar masses, estimated to have a surface temperature of 30,000 K and luminosity about 190,000 times the intrinsic brightness of our Sun. Its diameter is 17 times wider than the Sun, indicating a volume 4,913 times larger.

The secondary star in the primary binary δ Ori Aa2 has luminosity of 16,000 Suns, mass nine times solar, and surface temperature of about 26,000 K. Diameter of this type B1 main sequence star is seven times the solar width and its volume is correspondingly 343 times larger.

MINTAKA - ADDITIONAL DATA:

Catalogue Designations: Delta Orionis, δ Ori, 34 Orionis, HR 1852, HD 36486, SAO 132220, HIP 25930
****

Parallax: 2.6 mas
Distance: 380 parsecs, 1260 light years
****

Primary components: δ Ori Aa1 / δ Ori Aa2
****

Combined Apparent magnitude: 2.4 (2.5 / 3.9)
Absolute magnitude: -5.4 / -2.9
Luminosities: 190,000 / 16,000 solar

Luminosity classes: bright giant II / main sequence V
****

Apparent color blends: blue-white / blue-white
Peak colors: blue-white / blue-white
Color indices: (B-V): -0.31 / -0.39
Spectral types: O9.5 / B1
Surface temperatures: 30,000 / 26,000 K
****

Diameters: 17 / 7 solar
Masses: 24 / 9 solar
****

Coordinates: (2000.0) 05h32m -00°18' / 83.00 -0.30
Transit at 9 p.m. January 29
Proper motion: +1 -1 mas/year
Radial velocity: +19 km/sec.
****

Alnilam December 10

East **East-Southeast** **Southeast**

Orion

December – March

AL-neh-lam

1.7 mag. blue-white supergiant; 2,000 light years

Anchor Point Star, B0 Ia

****

Relative to the Sun, if it was the size of a baseball, Alnilam would then be 5.8 feet wide – one-third the width of a pitcher's mound. This model star would have to be about 623,000 miles from the baseball/Sun. That's equal to going around Earth 25 times; or about 2.6 times the Moon's actual distance from Earth!

If the Sun instead is visualized with the diameter of a Soccer ball, a model of Alnilam at this scale would be 5.3 meters wide – almost a third the size of a soccer field's center circle. Scaled distance from the soccer ball/Sun to our Alnilam model would be almost three million kilometers, an impressive distance equal to seven and three-quarter times the span between Earth and Moon.

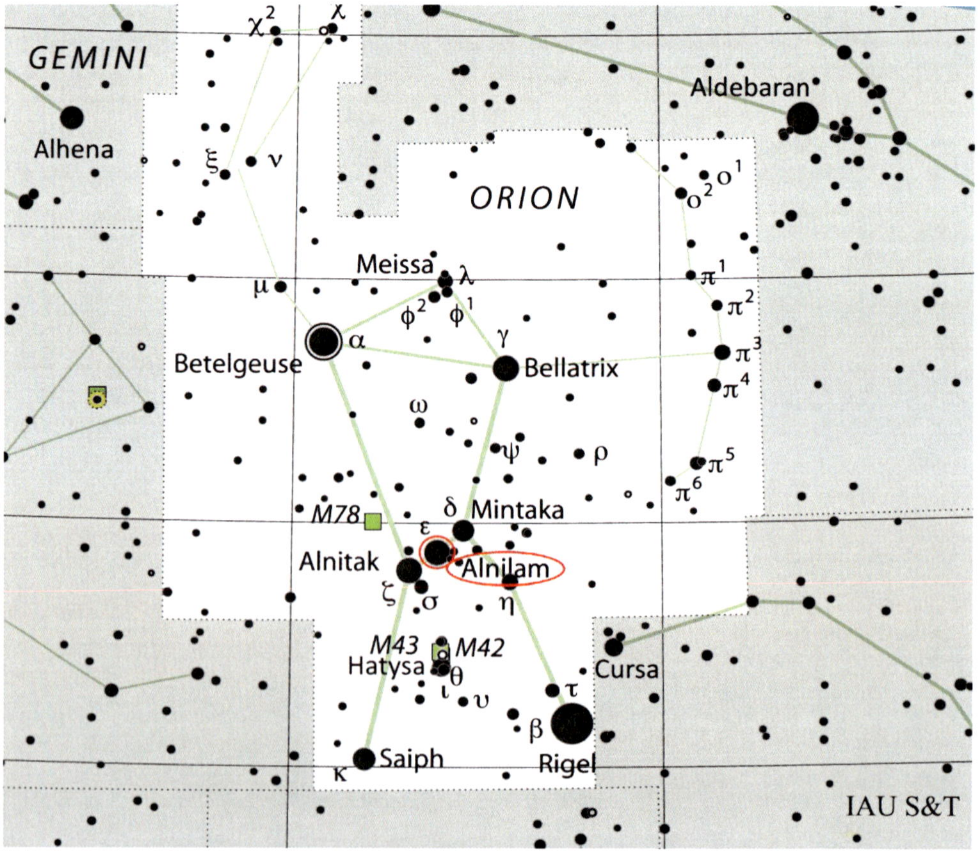

LOCATION:

Alnilam is the center star of the three forming Orion's Belt.

MEANING:

The name originated with an Arabic description of the Belt stars representing a "String of Pearls".

LORE:

The glorious symmetry of Orion's bright stars centers on Alnilam, a "Buckle" on the "Hunter's Belt". Some Biblical associations with those stars include "Jacob's Staff", "Peter's Staff", the "Three Magi", and "Three Kings". In Scandinavian sky traditions, Mintaka, Alnitak, and Alnilam were called "Freya's Staff", used by the Norse goddess of love. Greenland Eskimo folklore describes the three Belt stars as representing "Seal Hunters" who became lost at sea, then found their way into the heavens. To Australian aborigines, the stars are said to have shown "Three Young Men Dancing". The Pleiades were seven young maidens who played tunes for those celestial dancers. Chinook Indians of the Columbia River region of Oregon portrayed Belt and Sword of Orion as "Two Canoes".

HISTORY:

In 1964 Dutch astronomer Adriaan Blaauw (1917-2010) found that OB star groups such the Orion OB1 Association can devolve into sub-structures when individual stellar zones coalesce from gas and dust of the interstellar medium and start forming stars. He noted four sub-complexes of the Orion OB1 stellar association, each containing stars of different ages ranging from relatively young individuals in the Great Nebula M42, to others in adjacent stellar groups older than 10 million years.

Blaauw described one of those subsets, Orion OB1a, comprising stars sited northwest of the Belt. Another group, Orion OB1b, includes Belt stars Alnilam, Alnitak, and Mintaka, plus Sigma Orionis and others in their immediate vicinity. Stars in a set south of the Belt were listed as OB1c, and the oldest group OB1d encompasses stars in and around the Orion Nebula.

VISUAL ASPECTS:

Alnilam is one of the many blue-white supergiants in the Orion OB1 Association. These are massive stellar beacons of spectral types B and A which have rather recently evolved away from the main sequence. They are concentrated near the Orion molecular cloud complex, out of which they formed less than ten million years ago. The apparent magnitude of Alnilam is l.7, which, at this star's distance of about 2,000 light years, represents a luminosity about 70,000 times the Sun's. Supergiant Alnilam's spectral type B0 indicates a surface temperature of approximately 27,000 K

PHYSICAL NATURE:

The mass of Alnilam is 34.6 times that of the Sun and its diameter is 24 times greater than that of our nearest star. Therefore, the Sun is about 13,800 times smaller than supergiant Alnilam.

Alnilam serves as the standard spectrum for stars classified as B0 Ia in the Morgan, Keenan, Kellman Atlas of Stellar Spectra. The Atlas specifies the following criteria for a star's assignment to its B0 Ia supergiant category:

"The Si IV (silicon) absorption line at 4089Å wavelength shows a progressive decrease in intensity on passing from the very luminous supergiant ε Orionis (Alnilam) toward the main sequence star τ Sco. The He I(helium) lines 4387, 4144 and 4009 have a negative absolute magnitude effect and are strongest in the dwarf (main sequence star τ Sco)."

You might like to explore Caltech's interactive MKK Atlas, which shows criteria defining spectral type and luminosity class standards. It also has spectrogram images for various stellar categories. Our Appendix D lists of some important absorption lines referred to in that Atlas.

Supergiants such as Alnilam contain sufficient material so that even after episodes of mass ejection during their late evolution sufficient material remains to trigger core collapse supernovae explosions. Such destruction may induce formation of new generations of stars within surrounding nebulae, enriched by elements heavier than hydrogen and helium, forged in supernovae violence.

Shock waves from supernovae rip through molecular cloud complexes causing shear forces that induce sub-set cloud rotations amidst condensations of dust and gas. As pockets of interstellar matter contract and rotate at increasing speeds, they rise in temperature and pressure with some becoming protostars.

If those rotating condensations contain more than about 12 percent of the Sun's mass, further contraction raises their core temperatures to about 15 million Kelvins and substantial thermonuclear fusion of hydrogen may begin at their centers. With that happening objects are considered to have become main sequence stars. Succeeding generations of star formation may continue until all available interstellar material in an affected molecular cloud is consumed or dissipated.

Observations with McDonald Observatory's 16-foot radio telescope during the 1970's led to the discovery of a vast complex of star producing molecular clouds in Orion's sky sector.

ALNILAM - ADDITIONAL DATA:

Catalogue Designations: Epsilon Orionis, ε Ori, 46 Orionis, 112 G Orionis, HR 1903, HD 37128, SAO 132346, HIP 26311
****

Parallax: 2 mas
Distance: 600 parsecs, 2,000 light years
****
Apparent magnitude: 1.7
Absolute magnitude: -7.3
Luminosity: c. 70,000 solar
Luminosity class: supergiant Ia
****

Apparent color blend: blue-white
Peak color: blue-white
Color index (B-V): -0.10
Spectral type: B0
Surface temperature: 27,000 K
****
Diameter: 24 solar
Mass: 34.6 solar

****
Coordinates: (J2000.0) 05h36m -01°12' / 84.05-1.20
Transit at 9 p.m. January 30
Proper motion: +1 -1 mas/year
Radial velocity: +27 km/sec.
****

NEARBY FEATURES:

Orion Molecular Cloud Complex **– Observations of carbon monoxide made during the 1980's with Columbia University's 1.2-meter radio telescope extended initial discoveries of molecular clouds in and around Orion to other Galactic zones. The Columbia CO Survey was led by Patrick Thaddeus (1932-2017). That research continued after the telescope was dismantled from the roof of Pupin Hall of Physics, and he and the telescope moved to Harvard.**

The Orion Nebula M42 (images) represents a "hot spot" where stars currently are forming about 1,300 light years from the Solar System on the face of a vast system of molecular clouds. Along our line of sight behind the Orion Nebula

extend icy reaches of the Orion Molecular Cloud Complex. Those nebulae comprise mostly hydrogen and helium, but also contain many other types of gas as well as large quantities of interstellar molecular dust particles.

Temperatures of the clouds are estimated to average about 20 K, a level sufficiently cold to ensure that complex molecules survive, shielded by dust in the clouds from disruption by high energy ultraviolet radiation produced by neighboring stars. At its low temperature, the cloud's most abundant molecule hydrogen H_2, is not easy to detect. Instead, a surrogate in the form of molecular carbon monoxide produces a strong radio signal at 2.6-mm wavelength, enabling astronomers to trace outlines of the entire molecular cloud complex.

The Orion Nebula – Messier 42, at low magnification – asi

The vast molecular clouds of Orion range across this constellation from star Saiph northwest through the big hunter's "Sword", then northeast past Alnitak in Orion's Belt. Clouds have diameters of 100 to 200 light years and contain material equivalent to the masses of over 100,000 Suns. They contain much gas and dust expected someday to condense and form new stars.

As stars, dust, and gas revolve around the Galactic center, they periodically encounter regions defined by what are known as density waves, representing zones of compression. Therein stellar formation is stimulated, creating brilliant young stars and luminous gas clouds which define spiral arms in the Milky Way and countless other galaxies. This density wave theory of spiral arm formation was devised by Chia- Chiao Lin (1916-2013), and Frank Shu (1943-), in 1964.

In the Sun's neighborhood, active sites of current star formation activity are seen in the directions of constellations Taurus, Auriga, Ophiuchus, and most spectacularly, Orion.

Alhena December 12

East-Northeast East East-Southeast

Gemini

December – May

al-HEN-ah

1.9 mag. blue-white subgiant; 110 ly; spectroscopic binary

****

Imagine the Sun being the size of a baseball. On that scale Alhena would be about 10 inches wide at a scaled distance of about 34,270 miles from the baseball/Sun, equal to one and a third times a great circle trip around the world.

On the other hand, if the Sun had the diameter of a Soccer ball, another Alhena model at this scale would be 73 centimeters wide. Its distance from the soccer ball/Sun would be about 164,450 kilometers, that's about equal to a trip four times around the Earth, and in turn equivalent to 43% of the actual average distance from Earth to the Moon.

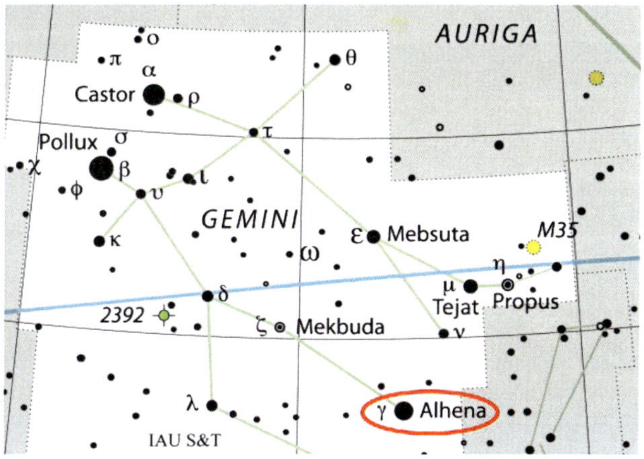

LOCATION:

Alhena appears about half way between Betelgeuse and Pollux, 14 degrees of arc northeast of the "red" supergiant Betelgeuse in Orion and 19 degrees southwest of exoplanet host Pollux, the more southerly of Gemini's twin stars.

MEANING:

The name Alhena was adapted from an Arabian description of it and several neighbors as a "Brand" on the neck of a celestial camel portrayed in this part of the sky.

LORE:

In ancient Babylon, Alhena and Eta Geminorum were described as the "Little Twins", comparing to Castor and Pollux, the "Twins of Gemini". Ptolemy identified Alhena as "Right Foot of the Eastern Twin".

VISUAL ASPECTS:

Alhena is the third-brightest star in Gemini and it shines with an apparent magnitude 1.9. It's a blue white subgiant whose luminosity is approximately 166 times that of the Sun. Alhena's parallax indicates a distance of 110 light years, and its spectrum suggests a surface temperature of about 9,300 K.

PHYSICAL NATURE:

This star system is a spectroscopic binary with an orbital period of 12.64 Earth years. The primary is a subgiant star with mass 2.8 times that of the Sun. Its luminosity is about 166 times the intrinsic brightness of the Sun. Alhena's mass is 2.8 times the Sun's mass and it is 3.3 times wider than the Sun. That equates to a 36 to one size difference between Alhena and the Sun. After seeing many similar comparisons to the Sun, it's not hard to understand why astronomers often describe "our" star as a dwarf.

Alhena's main sequence companion has a mass about 90% that of the Sun and a type G spectrum classification.

ALHENA - ADDITIONAL DATA:

Catalogue Designations: Gamma Geminorum, γ Gem, 24 Geminorum, HIP 31681, HD 47105, HR 2421, SAO 95912

****

Parallax: 30 mas

Distance: 33.3 parsecs, 110 light years

****

Apparent magnitude: 1.9

Absolute magnitude: -0.7

Luminosity: 166 solar

Luminosity class: subgiant IV
****
Apparent color blend: blue-white
Peak color: white
Color index (B-V): 0.00
Spectral type: A1.5
Surface temperature: 9,300 K
****
Diameter: 3.3 solar
Mass: 2.8 solar
****
Coordinates: (J2000.0) 06h38m +16°24' / 99.43 +16.40
Transit at 9 p.m. February 15
Proper motion: 14 -55 mas/year
Radial velocity: -13 km/sec.
****

NEARBY FEATURES:

Regions of sky just to the south and west of Alhena contain impressive sites producing new stars. Astronomers using infrared and radio telescopes ever since the 1970s have mapped vast molecular cloud complexes there, nebulae which provid raw materials for star formation activity considered second only to the Orion Nebula as a productive stellar nursery. In that zone, centered on the constellation Monoceros, the following sites are of special interest.

NGC 2264 comprising the Cone Nebula and Christmas Tree Cluster (06h41m +09°53') – These features, about six degrees of arc directly south of Alhena, also include a young star cluster with perhaps 1000 members and associated stellar birthing sites.

The cluster's brightest star, S Monocerotis, is a massive, highly luminous type O7 main sequence object with apparent magnitude 4.6. Several dozen young massive spectral type B main sequence stars also highlight the cluster. These objects are expected to be among the first to evolve and terminate as supernovae, providing another clue to NGC 2264's relatively young stellar age, about three million years. Additional support for that estimate is the presence of hundreds of protostars, still undergoing their gravitational contraction. These embryonic entities are too young to have reached the main sequence stage, when substantial proton-proton chain nuclear fusion will eventually occur in their cores.

NGC 2264 Cone Nebula, Christmas Tree Cluster , Fox Fur Nebula– asi

Notable examples of pre-main sequence objects in NGC 2264 include many protostars of the T Tauri variety as well as Herbig-Haro objects. At a distance of about 2,500 light years, NGC 2264 is considered second only to the Orion Nebula Cluster as a major star forming region within 3,000 light years from the Solar System. The cluster spans a sky area about 30 arcminutes wide, similar to that covered by the Moon. Stars of NGC 2264 represent the most significant component of the Monoceros OB1 stellar association. NGC 2264 is also associated with a nebula containing vast quantities of interstellar gas, glowing due to stimulation by ultraviolet light from that cluster's hot young stars. That nebula light contrasts well with an extensive dark and dusty molecular cloud background extending across a sky area about four times the apparent width of the Full Moon.

The Cone Nebula is a striking dark feature at the southern edge of NGC 2264. It's a shadow formed by the interstellar dust medium, concentrated by pressure from stellar winds, and outlined by light from S Monocerotis and neighboring stars.

The Christmas Tree Cluster is a sub-set of stars within NGC 2264 extending from the Cone Nebula north to S Monocerotis (the "tree" appears inverted). Also seen in its immediate area are the *Snowflake Cluster* and *Fox Fur Nebula*.

The introduction and increasing availability of digital imaging equipment and techniques during the 1990's allowed Astro-imagers to begin capturing unprecedented detail in their celestial "captures". That often lead to imaginative visualizations and descriptions of nebulae and star clusters previously considered unremarkable. Although the name Cone Nebula was used early in the 20th century, Christmas Tree Nebula, Snowflake Cluster, and Fox Fur Nebula became popular as designations after the Millennium.

Plaskett's Star in Monoceros, one of the most massive binary star systems known (6h37m+6°08') – is situated ten degrees south of Alhena. It's a spectroscopic binary estimated to be approximately 6,600 light years from Earth; and with apparent magnitude 6.1, it's near the threshold of naked-eye visibility. This stellar system's components are two of the most massive stars known, each estimated to contain about 55 times the Sun's matter. Individual luminosities are thought to be about 224,000 and 123,000 times that of the Sun. These mammoth components of Plaskett's Star have an orbital period of 14.4 days and appear to be surrounded by a considerable region of circumstellar dust.

Alnitak December 13

East **East-Southeast** **Southeast**

Orion

December – March

al-nih-TAK

1.9 mag. blue-white supergiant; 740 ly; multiple star system

****

Visualizing the Sun as the size of a baseball, at that scale Alnitak would be about nearly five feet wide – about one quarter the width of a pitcher's mound. That model would have to be about 230,500 miles from the baseball/Sun, a distance equal to nine and a quarter times the circumference of Earth and 97% of the Moon's average distance from Earth.

If the Sun was the size of a Soccer ball, an Alnitak model at this scale would be 4.4 meters wide – a quarter the size of a soccer field's center circle. Scaled distance from the soccer ball/Sun to our Alnitak model would be about 1,106,300 kilometers, nearly three times the distance between Earth and Moon.

LOCATION:

Alnitak is the southeastern star, at left end of the Orion Belt three.

MEANING:

Giuseppe Piazzi (1746-1826) conceived the name Alnitak for this star, whose Bayer designation is Zeta Orionis, at the beginning of the nineteenth century. He based the name on a transliteration of an Arabic word for belt. An alternate version, Mintaka was used by Piazzi to identify Delta Orionis at the right end of the Hunter's Belt.

LORE:

It is said sailors in the past referred to the Belt of Orion as the "Golden Yardarm". In France, the stars were identified as a celestial "Rake", and in parts of Germany the title "Three Mowers" was used. Alfred, Lord Tennyson (1809-1892), a devoted watcher of the skies, wrote these lines about the Belt stars of Orion:

"Those three stars of the airy Giant's zone, that glitter burnished by the frosty dark;"

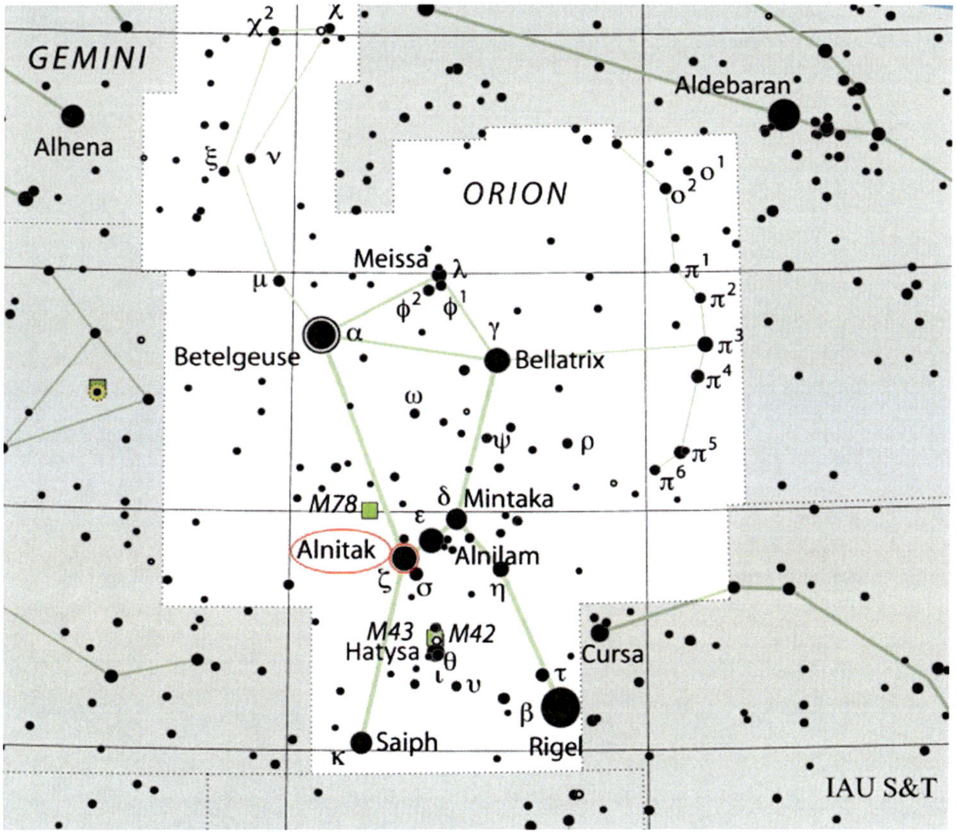

HISTORY:

During the early nineteenth century German amateur astronomer George Kunowsky (1786-1846) discovered that Alnitak has a fourth magnitude companion now known as Alnitak B. The primary itself was found to be spectroscopic binary as a result of observations conducted at Lowell Observatory during 1998. The close companion has designation Alnitak Ab, with Alnitak Aa being reserved for the primary, this system's brightest star.

VISUAL ASPECTS:

Primary star Aa in the Alnitak system is a blue-white supergiant of spectral type 09.7, with apparent magnitude 1.9. At visible wavelengths, this great star is 8,800 times more luminous than the Sun. It's among the most intrinsically bright stars in our sky; and its spectrum indicates a surface temperature of 28,000 K.

Visual secondary star Alnitak B has apparent magnitude 4.3. It's a blue-white giant at an angular distance of 2.4 arcseconds from the primary. Near the Alnitak triple system is a tenth-magnitude neighbor star at 58 arcseconds separation. Lack of observational evidence for orbital motion indicates it's probably only an "optical" neighbor, with no gravitational connection to the Alnitak system.

PHYSICAL NATURE:

Analysis of orbital characteristics in the spectroscopic binary system results in an estimated mass 33 times that of the Sun for the Alnitak primary component Aa. Diameter of this primary star in the Alnitak system is estimated to be 20 times the solar width. Therefore, this star is about 8,000 times larger than the Sun.

Stars of the Alnitak system are all members of the Orion OB I association and their copious outpourings of ultraviolet radiation ionized clouds of hydrogen in their vicinity. As a result, those nebulae glow softly at temperatures of about 10,000 K. Alnitak is surrounded by a lovely region of nebulosity, visible on photographs both by reflection of starlight to us from dust particles, and by the gas cloud's fluorescence.

ALNITAK - ADDITIONAL DATA:

Catalogue Designations: Zeta Orionis, ζ Ori, 50 Orionis, 126 G. Ori, HR 1948/9, HD 37742, SAO 132444, HIP 26727

****

Parallax: 4.43 mas
Distance: 226 parsecs, 740 light years

****

Apparent magnitude: 1.9
Absolute magnitude: -4.9
Luminosity: 8,800 solar
Luminosity class: supergiant Ib

****

Apparent color blend: blue-white
Peak color: blue
Color index (B-V): -0.11
Spectral type: O9.7
Surface temperature: 28,000 K

****

Diameter: 20 solar
Mass: 33 solar

****

Coordinates: (J2000.0) 05h41m -01°57' / 85.19 -1.94
Transit at 9 p.m. January 31
Proper motion: +1 +0 mas/year
Radial velocity: +18 km/sec.

****

NEARBY FEATURE:

The Horsehead Nebula in Orion (5h41m -2°28') – This wonderfully shaped figure in the form of a horse's head is less than one angular degree south of Alnitak, (seen at the top of this image, with a cross). The nebula is viewed on photographs as a jet of unilluminated dust which, at its distance of about 1,400 light years, partly blocks our view of the background bright nebula near Alnitak.

The Horsehead Nebula; directly south of Alnitak,
whose ultraviolet light causes adjacent Flame Nebula to glow – asi

This unique head-like shape was discovered in 1888 on a glass photographic plate, part of the remarkable Harvard College Observatory collection, by Scottish-American astronomer Williamina Fleming (1857-1911).

Star Notes is a Zooniverse project begun in 2020 engaging volunteers to help transcribe early notebooks of renowned Harvard astronomers including Ms. Fleming, Annie Jump Cannon, and Henrietta Swan Leavitt. Revealed references to image locations in the plate collection will enable cross-referencing, thereby enhancing access to these valuable records.

Amateur astronomy icon Walter Scott Houston (1912-1993), writing in the January 1970 issue of Sky & Telescope magazine, reported glimpsing the Horsehead (Barnard 33) from Connecticut immediately after passage of a strong cold front.

For that sighting, Houston used a 5-inch "Moonwatch" telescope, an instrument designed to be used by network of observers organized by the Smithsonian Astrophysical Observatory in 1956, in connection with the International Geophysical Year (IGY).

The intent of that project was to accurately record times and positions of the first generation of artificial satellites orbiting over teams of what now would be called citizen scientists. The operation was innovated by SAO director Fred Whipple and administered by Josef Allen Hynek. Over 200 teams of what we would now call "citizen scientists" used small refractors to track artificial satellites including those of the Soviet Union after Sputnik 1 was launched on the seminal day October 4, 1957, when the Space Age began.

Hynek is best remembered for his extensive work with UFO studies, including *Project Blue Book* and a cameo appearance in the classic Steven Spielberg film *Close Encounters of the Third Kind,* whose title *was* inspired by a category Hynek devised to describe high level contact with extra-terrestrial beings.

Using his compact Moonwatch telescope Houston related several successful sightings of the Horsehead Nebula by his readers, noting their observations were typically and not surprisingly helped by telescopes having superior optics and objective mirrors over ten inches in diameter. Also, viewing from sites at altitudes of several thousand feet or more under exceptionally clear skies without disturbing effects of wind was very beneficial.

"Scotty" as he was known in the astronomy community, also reported inability to spot the Horsehead by several other observers using – a 2.4-inch refractor, a 6-inch refractor with an f/8 focal ratio, and a 12.5-inch reflector at f/6.5.

The Horsehead, although a familiar sight in print and screen images, remains a considerable challenge when seeking to see it in person. Not only is a suitable telescope required, but successful observations greatly benefit from a high-altitude viewing site, a recently concluded wind storm, and rain-washed atmosphere.

Horsehead Nebula infrared image Credit: NASA, ESA, and the Hubble Heritage Team (STScI/AURA)

Rigel December 14

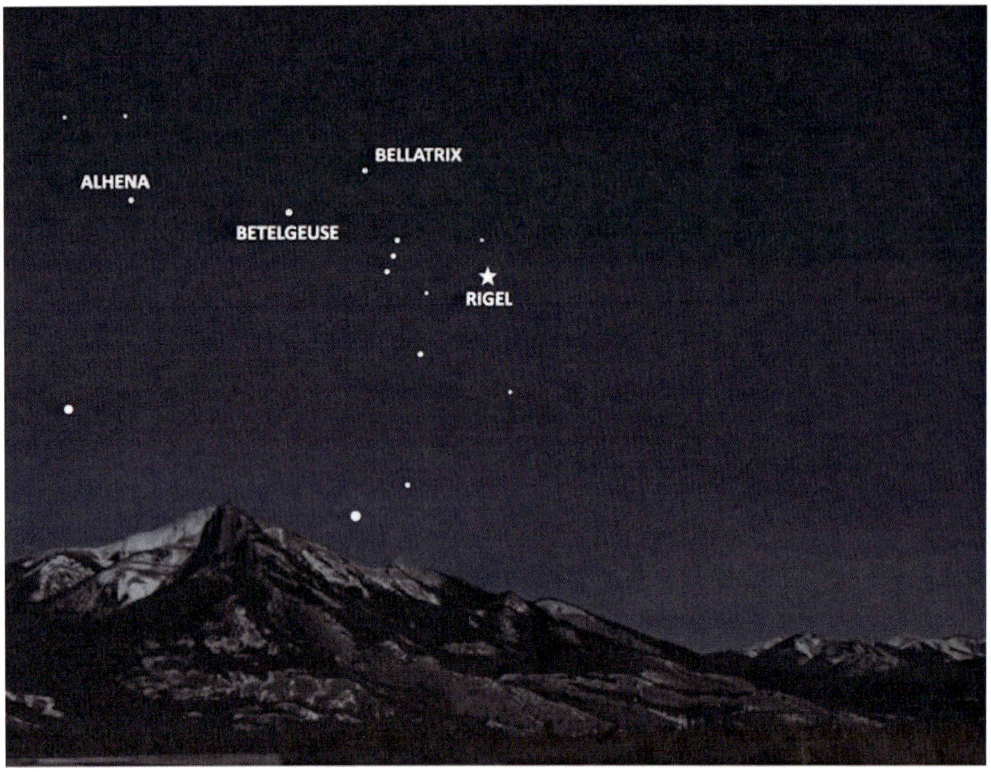

East-Southeast Southeast South-Southeast

Orion

December – March

RYE-jel

0.1 mag. blue-white supergiant; 860 ly; binary

Anchor Point Star – B8 Ia

****

Imagine the Sun being the size of a baseball. On that scale, Rigel would be about 19 feet wide – about the width of a pitcher's mound. A Rigel model of that size would have to be about 267,900 miles from the baseball/Sun, 112% of the Moon's actual distance from Earth.

If the Sun instead had the diameter of a Soccer ball, a Rigel model, at this scale, would be 17.4 meters wide – nearly the size of a soccer field's center circle. Scaled distance from the soccer ball/Sun to our Rigel model would then be about 1,285,700 kilometers, three and a third times the actual average distance between Earth and Moon.

LOCATION:

Rigel is the fifth-brightest star in our nighttime sky. It shines with blue-white brilliance nine degrees southwest of Orion's Belt.

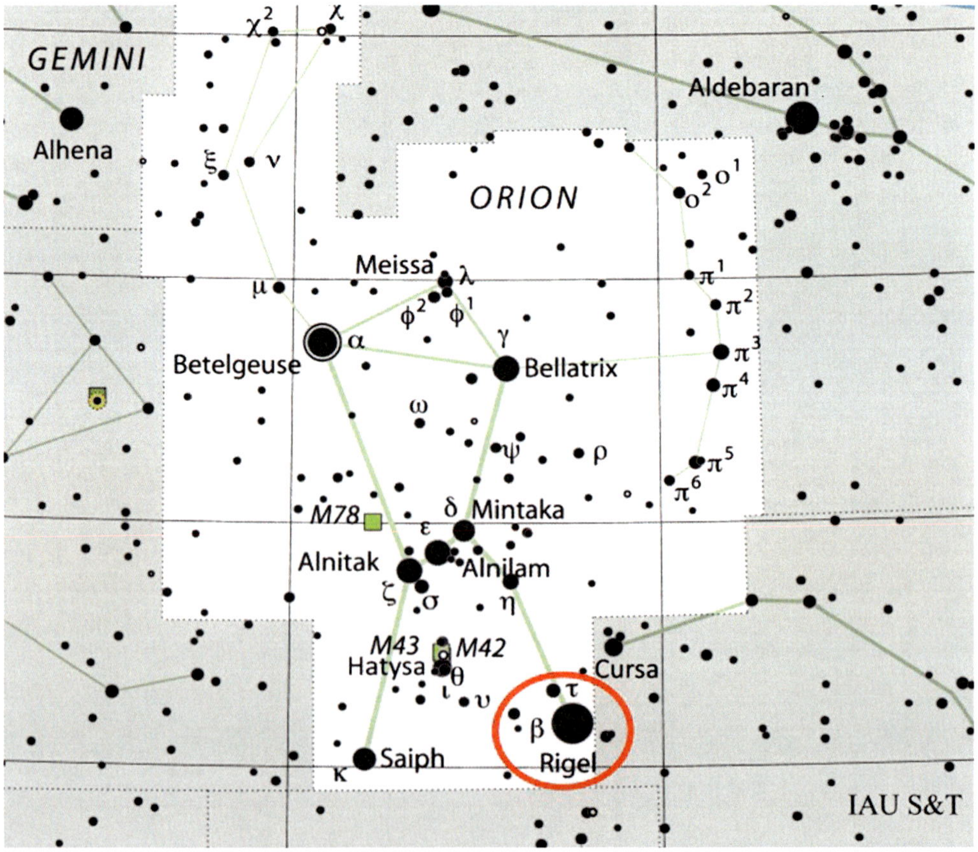

MEANING:

The name Rigel is taken from Arabian words for the "Left Foot of Orion". Paul Kunitzsch (1930 – 2020) notes it's one of the few star names in continual use with nearly the same spelling since the tenth century.

LORE: **In Scandinavian celestial mythology, Rigel represented the "Big Toe" of the giant Orwandil. His other big toe was broken off by Thor after it became frostbitten. The thunder god is said to have thrown the toe into the northern sky where it became the star Alcor, companion to Mizar in the Big Dipper.**

HISTORY: **Rigel's seventh magnitude companion was identified by Sherburne W. Burnham (1838-1921) in 1871, using his 6-inch refracting telescope.**

VISUAL ASPECTS:

Rigel is one of the most luminous stars known, with intrinsic brightness about 54,000 times greater than that of the Sun. It's a blue-white supergiant of apparent visual magnitude 0.1.

If Rigel was seen in our sky at the same distance as Sirius, it would shine about 144 times more brilliantly than Venus. The distance of Rigel is about 860 light years.

The surface temperature is about 11,100 K. Doppler shifts reveal slight radial velocity wiggles with a period of about 10 days; suggesting pulsations of the star's photosphere occur during that interval.

The seventh magnitude blue-white companion star is at an apparent separation of 9.5 arcseconds from Rigel. Both have the same proper motions and they're probably binary companions with a long orbital period. The overpowering brightness of Rigel makes it difficult to see this companion.

Rigel is another member of the Orion OB1 Association, a bright product of star formation in the constellation's molecular cloud complex.

PHYSICAL NATURE:

Rigel has 21 times the Sun's mass and 79 times the diameter of our nearest star. That indicates an approximate size 9,261 times larger than the Sun.

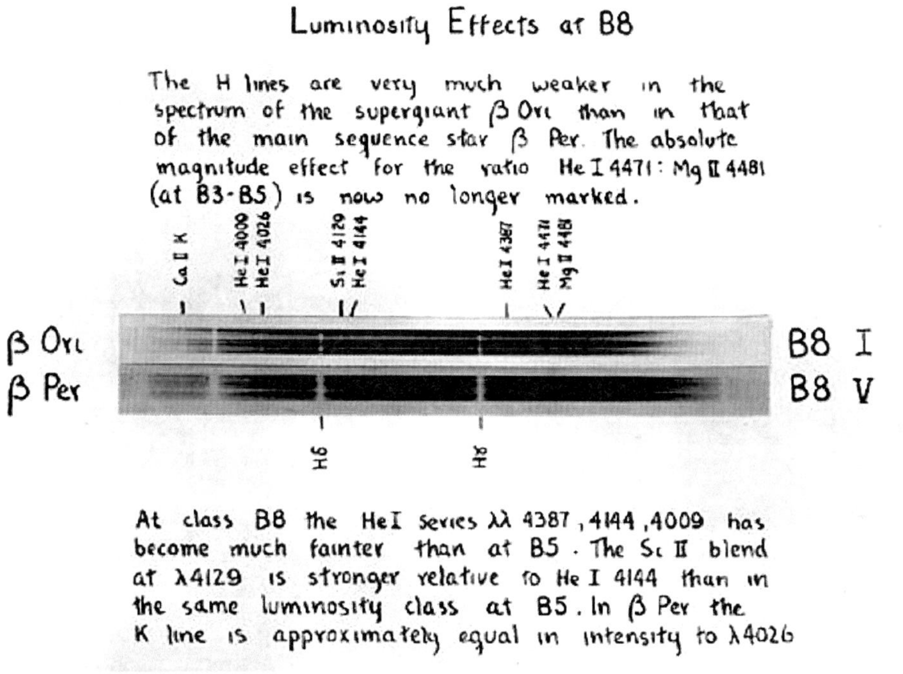

Rigel, with spectral type B8 and luminosity class Ia, is an important Anchor Point standard star in the MKK Atlas.

Therein, it's noted the hydrogen δ delta and hydrogen γ gamma absorption lines are very much weaker in Rigel's spectrum than they are in the comparison spectrum of main sequence B8 type β Persei (Algol). Also indicated that in other less luminous spectral type B8 stars, the selenium Se II blend at 4129 Ångstroms is stronger relative to the helium He I line at 4144 Å, seen in Rigel's luminosity class Ia spectrum.

Credit: MKK Atlas of Stellar Spectra

Classification of stars on the basis of such empirical comparisons benefit from being quite unaffected by differences between types of telescopes, instruments used to obtain data, and vagaries of observing conditions. Those are main reasons the MKK system and its later extensions continue in general use.

RIGEL - ADDITIONAL DATA:

Catalogue Designations: Beta Orionis, β Ori, 19 Orionis, HD 34085, HR 1713, HIP 24436, SAO 131907

****

Parallax: 3.8 mas
Distance: 263 parsecs, 860 light years

****

Apparent magnitude: 0.1
Absolute magnitude: -6.2
Luminosity: 54,000 solar
Luminosity class: supergiant Ia

****

Apparent color blend: blue-white
Peak color: blue-white
Color index (B-V): -0.03
Spectral type: B8
Surface temperature: 11,100 K

****

Diameter: 79 solar
Mass: 21 solar

****

Coordinates: (J2000.0) 5h15m -8°12' / 78.63 -8.20
Transit at 9 p.m. January 25
Proper motion: +1 +1 mas/year
Radial velocity: +18 km/sec.

****

NEARBY FEATURE:

Geminids Meteor Shower around December 14th **– This event reaches its peak intensity about December 14 and it's usually one of the best meteor displays of the year. Geminids meteors radiate across the sky from a point near the star Castor in Gemini, hence their name. It has been suggested the asteroid 1566 Icarus might be the worn-out nucleus of an extinct comet that once spawned Geminid meteors. Observers can expect to see about fifty trails per hour from this shower, with meteors plunging through the atmosphere at speeds averaging 35 kilometers per second.**

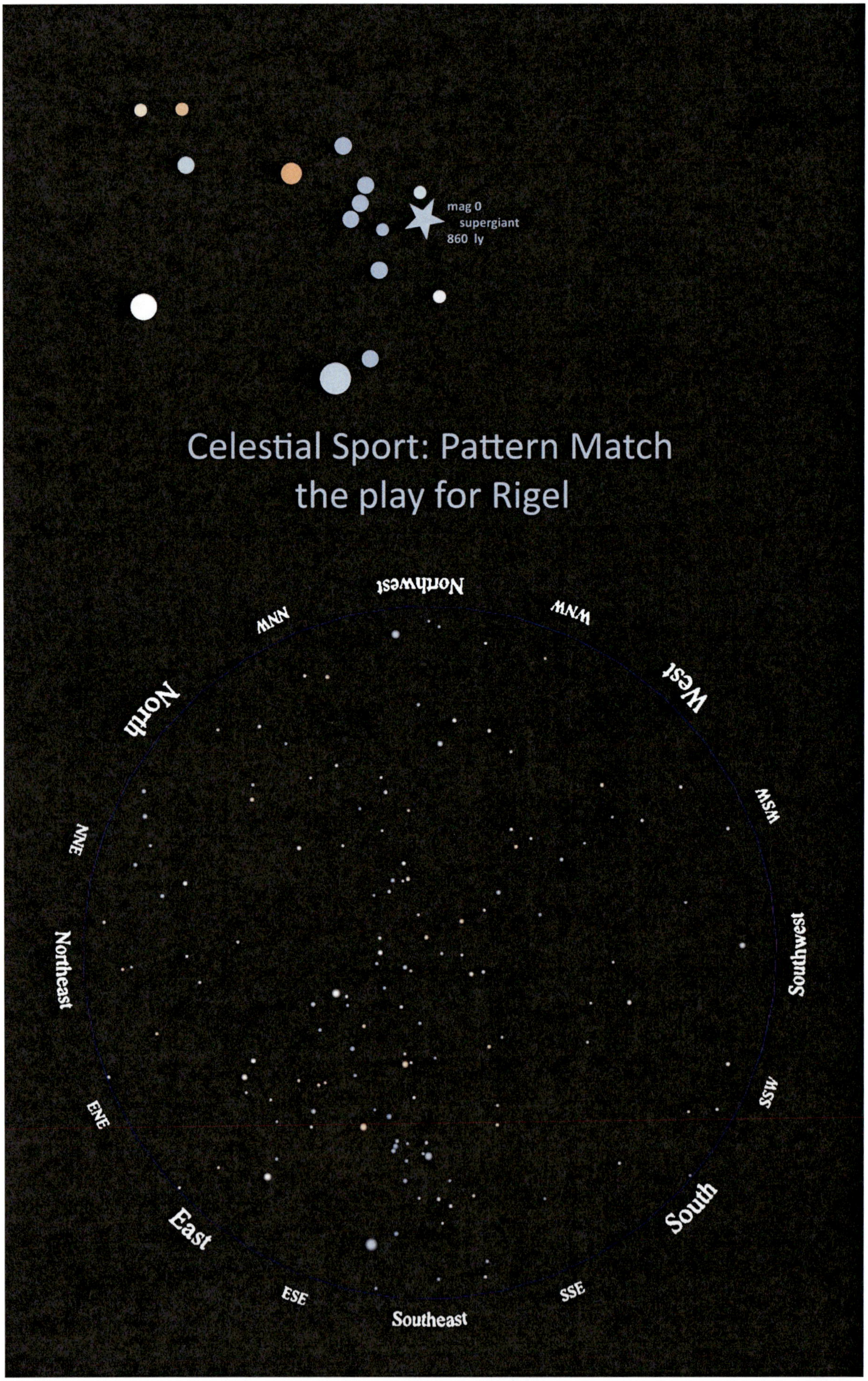
mag 0
supergiant
860 ly
Celestial Sport: Pattern Match
the play for Rigel
Northwest
NNW
WNW
North
West
WSW
NNE
Northeast
Southwest
ENE
SSW
East
South
ESE
SSE
Southeast

Hatysa December 16

East-Southeast **Southeast** **South-Southeast**

Orion

December – March

HA-tee-sah

2.8 mag. blue-white giant; 2,300 ly; spectroscopic binary primary in a visible triple star system

****

Envision the Sun as the size of a baseball. On that scale Hatysa would be about six feet wide – one-third the width of a pitcher's mound. That model would have to be about 716,500 miles from the baseball/Sun. That distance span would equal 29 trips around the world, or three times the Moon's actual distance from Earth!

If the Sun was the same diameter as a soccer ball, a Hatysa model, at this scale, would be five and a half meters wide – one third the size of a soccer pitch center circle. Scaled distance from the soccer ball/Sun to our Hatysa model would be about 3,438,500 kilometers. That metric distance is equal to 86 times Earth circumference, or nine times the distance from Earth to the Moon!

LOCATION:

Hatysa is a third-magnitude multiple star system in the "Sword" region of Orion, about five degrees south of Alnilam.

MEANING:

This star and objects in its immediate apparent vicinity were known to Greek astronomers as marking the "Sword of Orion". This description was later adopted in Arabia, where Hatysa was called the "Bright One of the Sword".

Earliest apparent use of the now official IAU name Hatysa was in a popular mid-twentieth century star atlas known as *Atlas Coeli* produced by Antonín Bečvář, who founded Skalnaté Pleso Observatory in what is now Slovakia.

Respected stellar etymologist Paul Kunitzsch could cite no prior source for this name.

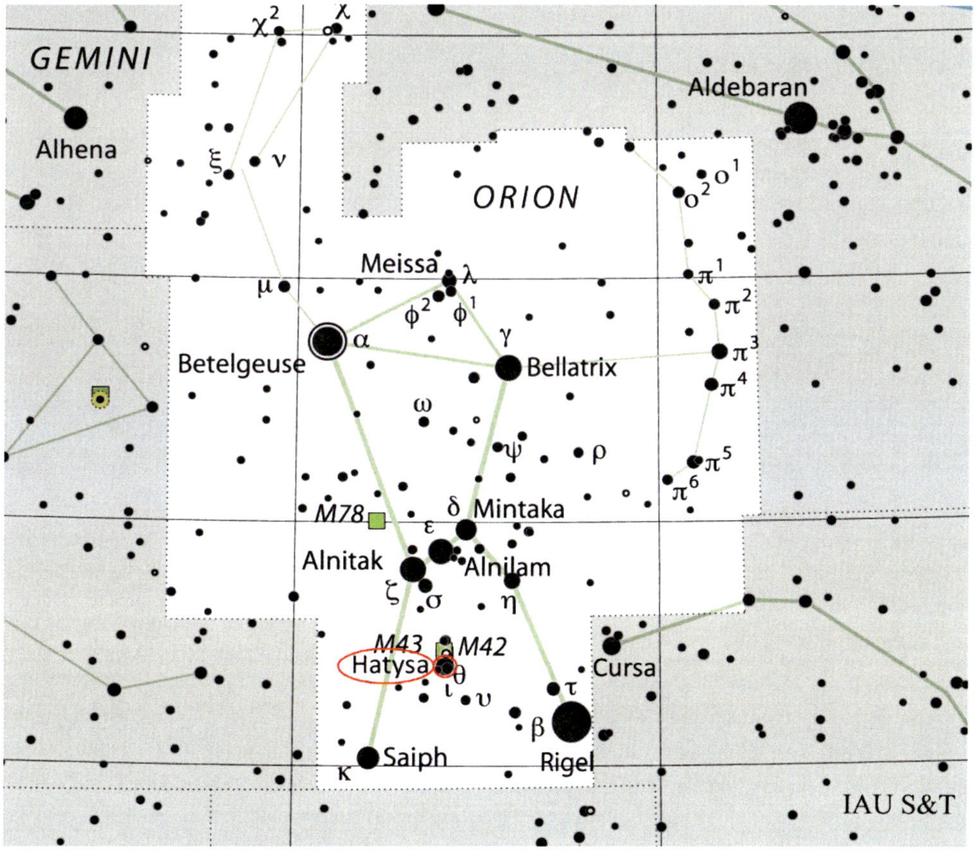

HISTORY:

Wilhelm Struve (1793-1864), the German director of Pulkovo Observatory near St. Petersburg Russia, in 1831 discovered Hatysa is a triple star.

VISUAL DESCRIPTION:

The three visible components of the Hatysa stellar system have shown little change in relative position since observed by Struve in the 19th century. Components of a primary spectroscopic pair, designated Hatysa A, have an orbital period of 29.14 days. A companion known as Hatysa B is a blue white subgiant of magnitude 7.3, with a separation of 11.4 arcseconds from its neighbor, the primary spectroscopic pair.

The visibly separate AB pair is an attractive sight for observers with telescopes of even a few inches in diameter. A third component star has magnitude 11, with a separation of about 50 arcseconds from the primary A.

The Hatysa system is at the fringe of the Orion Nebula, and its extremely hot young stars produce ultraviolet radiation that stimulates hydrogen and other gases in the Nebula to glow brightly enough to be seen with unaided eyes on clear, dark nights.

PHYSICAL NATURE:

Primary binary components of Hatysa, which we see as a single star, have a combined apparent magnitude of 2.8. This spectroscopic pair comprises an O9 blue-white giant about 34,000 times more luminous than the Sun and a spectral type B1 blue-white giant companion having luminosity 31,000 times that of the Sun. These two closely orbiting stars have surface temperatures estimated to be 32,000 K and 22,000 K respectively. In total, the Hatysa system comprises four massive young stellar members of the Orion OB1 Association.

The primary has 28 times the mass and 25 times the diameter of the Sun. Volume of this Iota Ori component A therefore is 15,625 times larger than the Sun. The second Hipparcos data reduction indicates a distance between Earth and the Hatysa system of approximately 2,300 light-years, making it one of the most distant stellar objects we can see with unaided eyes.

HATYSA - ADDITIONAL DATA:

Catalogue Designations: Iota Orionis, ι Ori, 44 Ori, HD 37043, HIP 26241, HR 1899, SAO 132323, A1216
****

Parallax: 1.4 mas
Distance: 714 parsecs, 2,300 light years
****
Apparent magnitude: 2.8 combined
Absolute magnitude: -6.5 / -6.4
Luminosities: 34,000 / 31,000 solar
Luminosity classes: giant III / giant III
****

Apparent color blend: blue-white / blue-white
Peak color: blue / blue-white
Color index (B-V): -0.27 / -0.20
Spectral types: O9 / B1
Surface temperatures: 32,000 K / 22,000 K
Diameters: 25 / 35 solar

Masses: 28 / 18 solar
****
Coordinates: (J2000.0) 05h35m 05°55'/ 75.58 -5.92
Transit at 9 p.m. January 30
Proper motion: +1 +0 mas/year
Radial velocity: +22 km/sec.
****

NEARBY FEATURES:

Theta Orionis – The Trapezium (5h35m -5°23') – Theta Orionis is a multiple star system in the heart of the Orion nebula, about half a degree of arc, one apparent Moon diameter north of Hatysa. The Theta system consists of four stars with a magnitude range of 5.1 to 6.7, and average separations of about 14 arcseconds. These four are easily visible to observers using small telescopes and received their nickname from the geometric pattern they form.

The entire stellar system of Theta Orionis is believed to contain at least twelve components, all relatively young formed from gas and dust of the Orion Nebula. Ultraviolet energy from these luminous stars helps make the Nebula itself glow by stimulating visible light fluorescence in ambient gas clouds.

M42: *The Great Nebula in Orion* (5h 35m -5° 27') – This is considered by many telescopic observers to be the loveliest object of the deep sky.

Methods of infrared and radio astronomy are especially useful in research regarding the Orion Nebula. As a result, it has been shown the Nebula sits like a hot button in front of a vast complex of molecular clouds stretching across most of Orion. The Great Nebula has a distance of approximately 1,600 light-years from Earth, with a diameter of about 30 light years.

To the naked eye, this Nebula resembles a hazy star. It evidently was not mentioned in pre-telescopic records, even by renowned observers including Hipparchus, Ulug-Beg, and Tycho Brahe. Nicholas-Claude Peiresc (1580-1637) is credited with first recorded mention of Orion's Great Nebula 1611. By 1656, Christiaan Huygens (1629-1695) of the Netherlands described it and the remarkable quadruple Theta Orionis seen at its heart. Charles Messier's catalogue of nebulae and star clusters contained a description of "the beautiful nebula of the Sword of Orion." Its position was measured by Messier in 1769, and sectors of the Nebula were assigned numbers 42 and 43 in his catalogue.

Across the English Channel, William Herschel (1738-1822) made the Orion Nebula the first observation of record whenever he placed in service each new instrument in the progression of ever-larger telescopes he constructed. Herschel tested each instrument by making detailed examinations of this marvelous object.

In 1825, William's son John Herschel (1792-1871) also an astronomer of the first rank and a pioneer in the field of photography who actually coined that name, described the Orion nebula with these words:

"I do not know how to describe it better than by comparing it to a curdling liquid,
or to the breaking up of a mackerel sky when the clouds of which it consists begin to assume a cirrus appearance."

Tennyson (1809-1892) wrote this about the Belt, Sword, and Nebula of Orion:

"A single misty star which is the second in a line of stars that seem a sword beneath a belt of three,
I never gazed upon it but I dreamt of some vast charm concluded in that star to make fame nothing ..."

Introduction of photography and spectroscopy to astronomy starting in middle decades of the nineteenth century revealed marvelous details of the Orion Nebula beyond the province of even the best unaided eye observers.

On a September night in 1880, Henry Draper (1837-1882) and Mary Anna Palmer Draper (1839-1914) used the 11-inch Clark refractor at their Hudson Valley observatory to make a 51-minute photograph of the Great Nebula in Orion. Henry's brother Daniel Draper (1841-1931) was also deeply involved with science. He established the New York Meteorological Observatory in Central Park's Arsenal building and served as the City's official meteorologist for 42 years

M42 the Orion Nebula above M43 – asi

In 1883, Englishman Andrew A. Common (1841-1903) achieved a milestone in astronomical imaging history by photographing details of the Orion Nebula invisible to the naked eye.

About that same time, William Huggins (1824-1910) and Margaret Lindsay Huggins (1848-1915) succeeded in photographing bright emission lines in Orion Nebula spectra from their back yard observatory near London. They thereby discovered that much of the Nebula consists of brightly glowing hydrogen and other gases. Since then, astronomers have applied myriads of new instruments and techniques to explore space in and around this remarkable object.

Alan Hirshfeld of the University of Massachusetts Dartmouth, in his excellent 2014 book, *Starlight Detectives*, details personalities and extraordinary contributions of individuals working alone or in small teams innovating and applying new technology as they practiced astronomy during the 19th and early 20th centuries

In 1946, Bart J. Bok (1906-1983) of Steward Observatory at the University of Arizona brought renewed attention to a number of roundish dark nebulae discovered many decades earlier by E. E. Barnard. Bok suggested the objects are regions of condensation, part of a process eventually leading to the formation of protostars in sky regions such those as in and around Orion.

Those nebulae, called Bok globules, vary in diameter, the smallest being about a half light-year across and the largest having diameters of several light years. Bok globules differ from other dark objects seen in certain nebulae, such as the Rosette in Monoceros. The Rosette Nebula's spots are believed, by some astronomers, to be "elephant trunk" structures as seen looking directly at their tip ends.

The Orion Nebula is a region where star production began probably less than a million years ago. Stars of the Trapezium system which illuminate Messier 42 are believed to have been formed well within that time frame. Ultraviolet energy from hot young Trapezium stars may be contributing to a second generation of stellar birthing at a site just behind the Orion Nebula in the adjacent molecular cloud.

Trapezium stars generate a region of hot expanding ionized hydrogen called a Stromgren sphere, which, as it pushes into the molecular cloud background, generates shock waves which may have triggered formation of a young infrared stellar cluster. The hypothesis was devised by Danish astronomer Bengt Stromgren (1908-1987)

Observations from the Kuiper Airborne Observatory (1974-1995) revealed an infrared protostar cluster within that background cloud, hidden by interstellar dust from the purview of optical telescopes.

Made hopelessly dim, as dust scatters optical light wavelengths, and situated just behind spectacularly visible portions of the Orion Nebula, is a region of new star production in that infrared cluster. The area is of considerable interest to astronomers. It contains stars estimated to be less than 100,000 years old – stellar babies on the cosmic time scale. They're so young they haven't had sufficient opportunity to shed most of their formative dust envelopes, and that masks the infant stars from our direct sight. However, as visible light photons from those nascent stars strike interstellar dust, some radiant energy is converted to heat wavelengths which then penetrates the dust and heads in our direction. After about 1,600 years travel time the heat is detected and imaged by means of high altitude as well as orbiting infrared telescopes such as JWST.

Combined Atacama Large Millimeter/submillimeter Array (ALMA), Gemini South and ESO Very Large Telescope image; Trapezium is at lower left of center; the BN Object is near center of the explosion flare.

In 1967, a point source of such infrared energy was discovered deep in the Orion Nebula. Known as the Becklin-Neugebauer object (BN) after its discoverers, Eric E. Becklin (1940-) and Gerald Neugebauer (1932-2014) of Cal Tech, it has long been described as a collapsing protostar, but Jonathan C. Tan at the University of Virginia suggested it's a full-fledged seven solar mass star that was dynamically slung out of its former binary orbit with huge 45 solar mass Trapezium star Theta^1 C (θ^1 C). That event may have happened "only" 4,000 years ago.

More recently it has been proposed to have been flung out by dynamic "dance" interaction with several other formerly close-orbiting protostars in an event that occurred about 500 years ago.

In any event, BN serves as an extensively researched example of a bright infrared source, providing clues as to how massive stars form from the interstellar medium. The "explosive flare" is only visible in infrared images.

In addition to inducement by shock waves from adjacent supernovae explosions, or intense ultraviolet pressure, another cause of star formation may be collisions and gravitational interactions between portions of giant molecular clouds. As such clouds encounter each other and move on; gas and dust at their interacting periphery may become compressed and sheared into rotation, triggering condensation to protostars.

Castor December 17

Northeast **East-Northeast** **East**

Gemini

December – May

CASS-ter

1.6 mag. blue-white main sequence; 49.2 ly; rare sextuplet star system

****

Imagine the Sun being the size of a baseball. On that scale Castor Aa, the brightest component of this six-star system, would be about seven inches wide. This model would have to be 15,900 miles from the baseball/Sun, equal to about two-thirds of the way around the entire world.

If the Sun had the diameter of a Soccer ball, a model of primary component Castor Aa at this scale would be 51 centimeters wide. Scaled distance from the soccer ball/Sun to our Castor model would be about 76,245 kilometers, approximately equal to a trip around the world twice.

LOCATION:

Castor is one of Gemini's famous Twin Stars. With second magnitude brightness it is located 21 degrees northeast of Alhena. If you sometimes find it difficult to distinguish between nearly identical twins, these two included, remember when in the middle of the sky they're stacked in alphabetical order, Castor "on top" crosses the meridian first, followed by "lower" Pollux. However, don't feel bad for Pollux; it's 58% brighter than Castor.

MEANING:

The name Castor was widely used to reference this star in ancient Greece and Rome, named after the mortal half-brother of Pollux, son of Zeus / Jupiter...

LORE:

Castor was known in Mesopotamia as the "Western One of the Twins", and described by Ptolemy as the "Star in the Head of the Western Twin". Arabian astronomers called Castor the "Head of the Foremost Twin". In classical mythology, Castor was the twin who was an expert trainer of horses, the mortal son of King Tyndareus and Queen Leda of Sparta. The name Apollo was also used for this star until around the mid-eighteenth century.

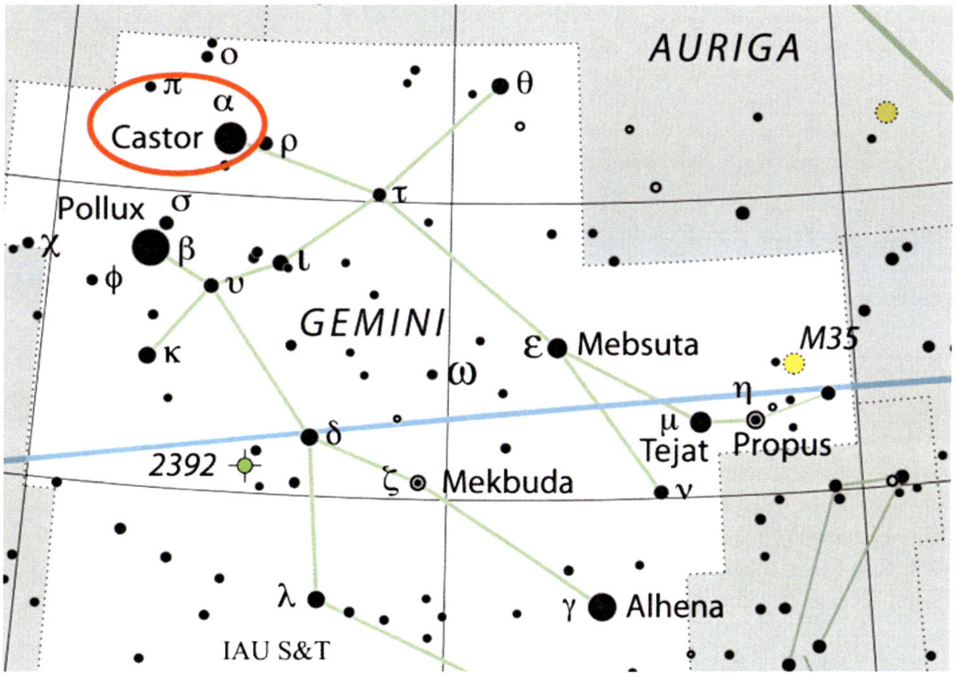

HISTORY:

The point of light we see with unaided eyes as the bright star Castor is actually a rare six-star system. Separation between its two brightest components may have been resolved by Italian astronomer Giovanni Domenico Cassini (1625-1712) as early as 1678 while he was first director of the Paris Observatory.

However, the binary nature of Castor was neither confirmed nor studied until 1719, when James Bradley (1693-1762) began observations, recording the two main component's positional changes over a period of years.

William Herschel later made detailed studies of Castor. By 1803, after component Castor B had completed about 58% of a complete orbit, Herschel became convinced Castor's AB main components move in gravitationally bound, elliptical orbits around their center of mass. That was a major discovery. It showed orbital motions of the Castor binary star system were in accord with Kepler's laws describing how planets orbit the Sun. Herschel's observations thereby provided first recorded evidence gravity acts in distant realms of space the same way it behaves affecting objects in the Solar System.

In 1897 Aristarkh Belopolsky (1854-1934), a stellar spectroscopy expert and later the director of Russia's Pulkovo Observatory, discovered Castor's secondary component B is a spectroscopic binary. Seven years later at Lick observatory American astronomer Heber Curtis (1872-1942) spectroscopically determined the binary status of Castor A.

Curtis is best remembered for his Great Debate with Harlow Shapley (1885-1972) contested in 1920 at the Smithsonian Institution's National Museum of Natural History. Shapley argued that galaxies, - "spiral nebulae" as they were then called, such as M31 in Andromeda and M101, the Pinwheel in Ursa Major, were relatively minor facets of the Milky Way Galaxy – in Shapley's mind the Milky Way was indeed the entire Universe. On the other side, Curtis believed what is now known to be true – those lovely intriguing spirals are in fact "island universes", distinct, vastly distant equivalents of the Milky Way.

Although notions of science are not ultimately decided by debate, the Shapley-Curtis event is memorable for having stimulated wide interest. We'll bet more than a few youngsters of that era were inspired to become astronomers, and went on to significantly contribute to our understanding of the universe, because they had been tuned in to the S-C encounter.

With a nod to that legacy, since 2001 the American Museum of Natural History annually sponsors the Isaac Asimov Memorial Debate; lively panel discussions moderated by Neil deGrasse Tyson (1958-) that unfold to enthusiastic audiences in packed auditoriums.

Rear L to R; Marcy Pedas Sigler, Steven L. Beyer, and co-hostess Andrea Elgin Beyer

Sing-a-long during gathering chez Sigler, credit & permission – Marcy Pedas Sigler

Neil deGrasse Tyson was 16 years old when he first met Isaac Asimov. They were among 2,600 solar eclipse hunters, including Neil Armstrong and Scott Carpenter, second American to orbit the earth, enjoying an Atlantic Ocean voyage. During the beautifully long five minutes 44 seconds of totality on the morning of June 30, 1974 they and their shipmates saw the eclipsed Sun among stars of the constellation Gemini, near Castor and Pollux, which formed a nearly straight line with brilliant Venus.

It was the world's second cruise ship eclipse hunting expedition. The first had sailed from New York Harbor the previous year, also innovated, organized and led by our friends the Pedas-Sigler family. Their entrepreneurial concept of cruising and bringing together eclipse spotters with telescopes, enhanced by cultural and educational programs, spawned an entirely new industry. Fast ship observing platforms enabled far better odds of dodging clouds and fine-tuning rendezvous positions featuring blue sky just before totality. Over the years their expedition model has provided many thousands with great pleasure and thrills by experiencing one of the most magnificent natural phenomenons.

For some aboard the expedition cruise ships, P&O line's *Canberra* and its partner *Cunard Adventurer*, it was a once in a lifetime experience. But many others were already planning to chase future solar blackout spectaculars including the Great North American solar eclipses of August 21, 2017 and April 8, 2024.

Castor C is the third spectroscopic binary in the Castor system. Its double nature was discovered and extensively described by Walter S. Adams (1876-1956) and Alfred H. Joy based on their research conducted during 1916 at Mt. Wilson Observatory. Adams was a pioneer in discerning critical relationships between relative strengths of various spectral lines, then applying them as correlates of luminosity classes such as main sequence, dwarf, and giant stars. He also discovered an abundance of carbon dioxide in the atmospheric infrared spectrum of Venus.

Alfred H. Joy (1882-1973) was an expert on stellar distances and the role of spectroscopic methods analyzing Doppler shifts of spectral lines. He also researched T-Tauri proto-stars, turbulently varying in appearance on the verge of igniting core hydrogen fusion, before graduating to full membership in the stellar community.

VISUAL DESCRIPTION:

There are three visible components in the Castor system, each an elusive spectroscopic binary, designated Castor A, Castor B, and Castor C.

Star systems containing six spectroscopic binary components are exceedingly rare, and few are known to contain an eclipsing binary such as Castor C.

Castor A and Castor B image are the visual pair revealed by Bradley. Component A has a combined apparent magnitude of 1.6, comprising a blue-white main sequence star and a red dwarf. With magnitude 2.9, component B is also a spectroscopic binary, as is the eighth magnitude C component pair. The National Optical Astronomy Observatory's Multiple Star Catalogue (click the link, "enter number of star" = 36850, and press "go" for Castor) provides details about this and other multiple star systems.

The second reduction of Hipparcos mission data was reported in a 2007 volume of the European journal *Astronomy and Astrophysics* by Floor van Leeuwen of the Institute of Astronomy, University of Cambridge. That analysis provided improved accuracy over the original Hipparcos results, and yielded a parallax of 64.12 (+/-3.75) milliarcseconds for the Castor star system. This corresponds to a distance of 15.6 parsecs, 50.8 light years, with a margin of error indicated as plus or minus three light years.

The Gaia *Archive* lists parallax of the Castor C pair as 66.23 mas, corresponding to 15.1 parsecs or 49.2 light years. In this case, the margin of error is only 0.051 mas (+/- 0.037 ly!)-Component C's magnitude is indicated by Gaia as 8.3.

For centuries, separation between Castor's brightest components could be resolved with a small telescope. However, by 1949 they reached a point in their 445-year orbit where angular distance between them reduced to less than three arcseconds; and minimum separation of 1.8 arcseconds occurred in 1970. During that era visual separation of components Castor A & B was a challenge for observers using small telescopes. Separation of these stars continues to increase, and Castor A B is again a fine looking double, with an angular separation of about 4.3 arcseconds in 2020.

PHYSICAL NATURE:

Castor Aa, brightest component of the sparkle we see with unaided eyes as Castor, is a spectral type A1 main sequence star with apparent magnitude 1.98 and mass 2.7 solar. Its diameter is 2.3 times that of the Sun. This primary component of the Castor system is slightly more than twelve times the size of the Sun.

The spectroscopic partner star, Castor Ab is a faint type M red dwarf with a mass only 0.21 solar. That's barely above the minimum 0.12 solar mass necessary for core fusion reactions to operate. Nevertheless, it is sufficient to separate this star from the realm of brown dwarfs. The Castor Ab red dwarf orbits Castor Aa once every 9.2 days.

Separated from the Castor A pair by a bit over four arcseconds is the B component noted by James Bradley. It comprises Castor Ba, a type A2 main sequence star with visual magnitude 2.88 and mass 1.82 solar; and component Bb. Castor Bb is another red dwarf. Its mass is 0.36 solar, barely big enough to be a star. However red dwarfs are very conservative in hydrogen fuel consumption and all are expected to endure for hundreds of billions of years! Red dwarf Castor Bb orbits Castor Ba every 2.9 days. Orbital period of the dual stars in Castor B around the two in Castor A is estimated at 445 years.

At a considerable span of 71 arcseconds, Castor C comprises Castor Ca, a spectral type M0.5 red dwarf, with 0.59 solar mass and a companion, Castor Cb, a nearly identical red dwarf of 0.58 solar mass. Their orbital period is 19 hours, 26 minutes. The Castor C spectroscopic pair's orbital period around the bright Castor A B set is estimated at about 15,000 years.

Faint Castor C, with combined magnitude listed by Gaia as 8.3, is also known as YY Geminorum – a rare eclipsing binary system. While both component C stars appear yellow-orange, their most intense individual color is the red typical of spectral type M stars, and their overall strongest radiation is infrared.

The orbital plane orientation of Castor Ca and Castor Cb, aimed directly toward Earth, causes periodic stellar eclipses enabling precise measurements. Thereby astronomers discovered the components of Castor C have approximately equal masses, about 60 percent that of the Sun. Component luminosities are about three percent of solar, with star diameters approximately 62 percent of the Sun's. Details about dimensions of the Castor C pair were provided in a 2002

Astrophysical Journal paper by Guillermo Torres and Ignasi Ribas, of the Center for Astrophysics, and the Department of Astronomy and Meteorology, University of Barcelona.

Analysis of the Castor C system is of high value to astronomers because Castor is one of only a handful multiple star systems containing binary low mass spectral type M main sequence stars. Details of individual stellar elements in this system contribute to greater understanding of Castor's individual components, the system as a whole, and the universe of red dwarfs.

CASTOR - ADDITIONAL DATA:

Catalogue Designations: Castor Aa, Alpha² Geminorum, α² Gem, 66 Gem, HD 60178, HIP 36850, HR 2891, SAO 60198

****

Parallax: 66.23 mas
Distance: 15.1 parsecs, 49.2 light years
****
Apparent magnitude: 1.6 (combined ABC)
Absolute magnitude: 0.9 (primary Aa):
Luminosity: 36 solar
Luminosity Class: main sequence

****
Apparent color blend: blue-white
Peak colors: white
Color index: (B-V): 0.03
Spectral type: A2
Surface temperature: 9,000 K
****
Diameter: 2.3 solar

Mass: 2.7 solar
****
Coordinates: (J2000.0) 07h35m + 31°53' / 113.75 +31.89
Transit at 9 p.m. March 1
Proper motion: 170 -102 mas/year
Radial velocity: -1 km/sec.
****

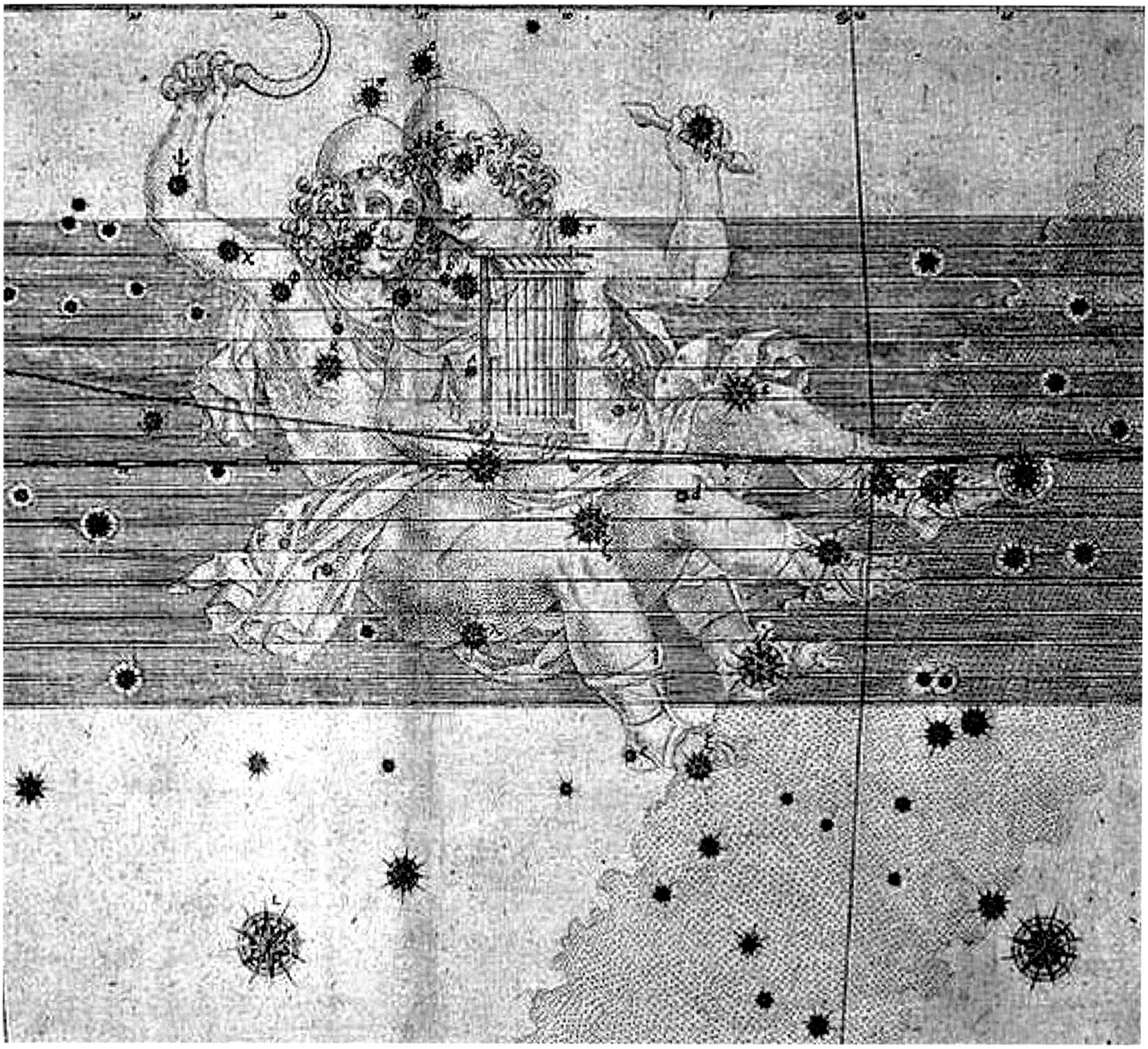

Gemini – Johann Bayer, Uranimetria 1603

Pollux December 21

Northeast **East-Northeast** **East**

Gemini

December – May

POL-lucks

1.1 mag. yellow giant; 34 ly; high proper motion star
Anchor Point Star – K0 IIIb
Brightest star hosting a confirmed exoplanet planet
****

If the Sun is imagined as the size of a baseball, at that scale Pollux would be about two feet wide. Our model Pollux would have to be placed about 10,600 miles from the baseball/Sun, approximately the distance between New York and northwestern Australia in order to be at the scaled distance.

Alternately, if the Sun had the diameter of a Soccer ball, a Pollux model would be almost two meters wide. Scaled distance from the soccer ball/Sun to our metric Pollux model would be about 50,800 kilometers, equivalent to a flight one and a quarter times around the world.

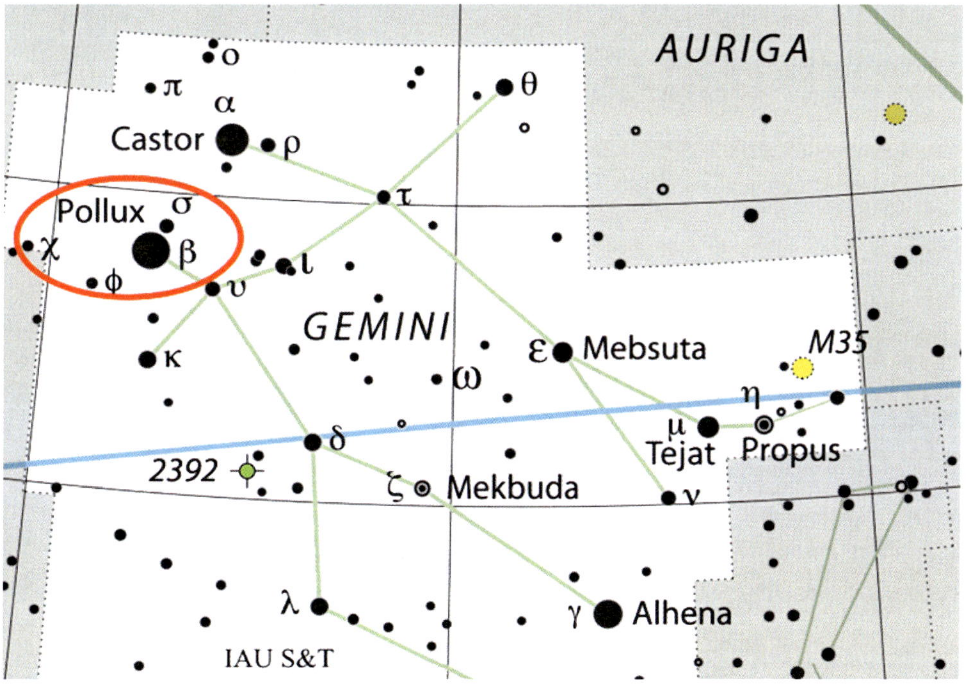

LOCATION:

Pollux is the brighter of the Twin Stars in Gemini. It is located four degrees southeast of Castor; and lower in the sky as it crosses the meridian after Castor.

MEANING:

Pollux represents the immortal Gemini twin, said to have been the son of Zeus and Queen Leda of Sparta. He was brother to Helen of Troy and half-brother of Castor.

LORE:

The "Twin" stars in Gemini form a visually striking pair; mutually associated throughout history in many cultures. In Babylon Pollux was called the "Eastern One of the Twins" and also "Yoke of the Enclosure". Ptolemy described Pollux in his *Syntaxis* as the "Red Star in the Head of the Eastern Twin"; an exaggeration of this star's yellowish color. In Arabia, Pollux was known as "Head of the Hindmost Twin". Pollux is characterized as hindmost because it transits the meridian about 10 minutes after Castor. Sometimes Pollux was named "Hercules", the companion of Apollo, as an alternate characterization in Hellenic astronomical lore.

HISTORY:

In a 2018 *Astronomy & Astrophysics* paper, astronomers led by Michel Auriere from the University of Toulouse reported research which provided first direct evidence of a weak magnetic field in the photosphere of Pollux. That group

also noted variations they observed in the star were consistent with published orbital properties of a recently discovered exoplanet whose existence was later confirmed, and named Thestias.

VISUAL ASPECTS:

Pollux is the twelfth brightest star in our night sky. With apparent magnitude 1.1, it's 63 percent brighter than its neighboring Twin Castor. Pollux is a yellow giant with a surface temperature of approximately 4,700 K, and it's about 31 times more luminous than the Sun. Although there are several faint stars apparently near Pollux in a telescope's field of view, their space motions indicate none are physically related to Pollux.

December 21st is a time of year when Pollux is ascending to prominence in the eastern sky. It's also a date when the Sun usually reaches its most southerly position on the ecliptic; winter begins in the Northern Hemisphere, and summer starts south of the equator.

At the moment northern winter begins, the center of the Sun's disk is on the 18-hour circle of right ascension and at -23.5 degrees declination. Our nearest star is then in a direction about one degree of arc northwest of M8, the Lagoon Nebula, just above the Teapot's Spout in Sagittarius.

Sunlight's footprint is longest with its least potency on the first day of winter. It is then most spread out, providing lowest energy concentration of the year. However, in many locales average daily temperatures continue to drop especially near coastlines due to slowly diminishing reserves of residual heat in the ground and water, lingering from warm autumn days. Into the new year cooling is increasingly compensated for by sluggish daily increases of sunlight concentration until (for example New York City) average daily temperatures start to rise around the 21st of January.

It is cheering to remember that when winter begins, the Sun ceases its apparent southerly motion and slowly but surely slides northeast along the ecliptic. Daylight duration times increase, sunlight's footprint becomes more concentrated, and springtime inevitable returns to the northern world. It's a cause for traditional year-end celebrations across the hemisphere.

PHYSICAL NATURE:

Trigonometric parallax of Pollux indicates a distance of 10.4 parsecs, equal to 34 light-years. That span is close to the standard distance of 32.6 light years used for calculations of absolute magnitudes. As a result, the apparent and absolute magnitudes of Pollux have nearly the same value. Pollux is the closest stellar giant to Earth and it's an MKK Anchor Point star, the standard for classification of all type K0 III class giants.

With twice the Sun's mass and a diameter 8.8 times wider than the Sun, Pollux is 680 times larger than the Sun

POLLUX - ADDITIONAL DATA:

Catalogue Designations: Beta Geminorum, β Gem, 78 Gem, HD 62509, HIP 37826, HR 2990, SAO 79666

Parallax: 96.5 mas
Distance: 10.4 parsecs, 34 light-years
****
Apparent magnitude: 1.14
Absolute magnitude: 1.1
Luminosity: 31 solar
Luminosity class: giant III
****

Apparent color blend: yellow
Peak color: yellow-orange
Color index (B-V): 1.00
Spectral type: K0
Surface temperature: 4,700 K
****
Diameter: 8.8 solar
Mass: 2 solar

****
Coordinates: (J2000.0) 07h45m +28°02' / 116.33 +28.03
Transit at 9 p.m. March 4
Proper motion: -627 -46 mas/year
Radial velocity: +3 km/sec.
****

NEARBY FEATURES:

Thestias Pollux's planet – The brighter twin of Gemini is orbited by a so-called super Jupiter, a world estimated to be about 2.9 times the mass of our Solar System's largest planet, and about 920 times the mass of Earth.

In 2014 officially named Thestias by the International Astronomical Union and also designated as HD 62509 b and Pollux b this big planet orbits about 1.69 AU from Pollux. Discovered by astronomer Artie P. Hatzes (1957-) in 2006, it revolves around Pollux once every 590 Earth days...

There are several other planets within sky borders of Gemini. However, they're a lot more elusive than Venus, Jupiter, Mars, Saturn, or even Mercury when they happen to be in the constellation of the Twins.

Nevertheless, we can imagine and visualize a bit when gazing at the general direction of those far more remote worlds while considering them. For example, after we've spotted Pollux on a frosty December night, we also remember Thestias is out there too. Distant planets can be visualized and savored in our mind's eye – imaginary, nonetheless marvelous to consider!

At least two other stars in Gemini beside Pollux are now known to host planets. Fifth magnitude orange giant HD 59686 (07h32m +17° 05') in the southern part of the constellation, 320 light years from us gravitationally tethers a planet having twice Earth's mass. That distant world orbits its star at a span about equal to that of Venus from the Sun, taking 303 days to complete a circuit.

Also, a seventh magnitude white main sequence star at 06h55min. +24°15', near Epsilon Geminorum, is orbited by planet HD 50554b. Ninety-eight light years from Earth, that world has the mass of five Jupiters – twice the heft of Pollux's Thestias, and orbits 2.4 astronomical units from its host star over a period of 1,300 days.

Saiph December 24

East-Southeast Southeast South-Southeast

Orion

December – March

SAFE

2.1 blue-white supergiant; 650 ly

****

Imagine the Sun being the size of a baseball. On that scale Saiph would be about five feet four inches wide – about a third the width of a pitcher's mound. That model would have to be about 202,500 miles from the baseball/Sun, equivalent to a trip about eight times around the world, and equal to 85% of the Moon's distance from Earth.

If the Sun had the diameter of a Soccer ball, a model of Saiph, at that scale, would be almost five meters wide – a quarter the size of a playing field's center circle. Scaled distance from the soccer ball/Sun to our Saiph model would be about 971,800 kilometers, two and a half times the distance from Earth to the Moon.

LOCATION:

Saiph is the last bright star of Orion to ascend onto the Sky Screen. It's a second-magnitude object, at the southeast corner of that constellation's figure – five degrees to the lower left of Hatysa and the Orion Nebula.

MEANING:

The name Saiph was applied to this star by Giuseppe Piazzi (1746-1826), borrowed by him from a translation of an Arabian description meaning "Sword of Orion".

LORE:

In antiquity, Ptolemy placed Saiph at the right (eastern) knee of Orion as the giant's image faces toward us. However, the constellation figure portrayed in Bayer's Uranometria of the early seventeenth century shows Orion facing away, with Saiph at the mighty Hunter's left knee. The rise of Saiph completes our view of the hourglass outline forming Orion's magnificent figure. This glorious constellation rises higher hourly and nightly to dominate winter evening skies, while its remarkable features captivate imaginations and intellects of sky watchers.

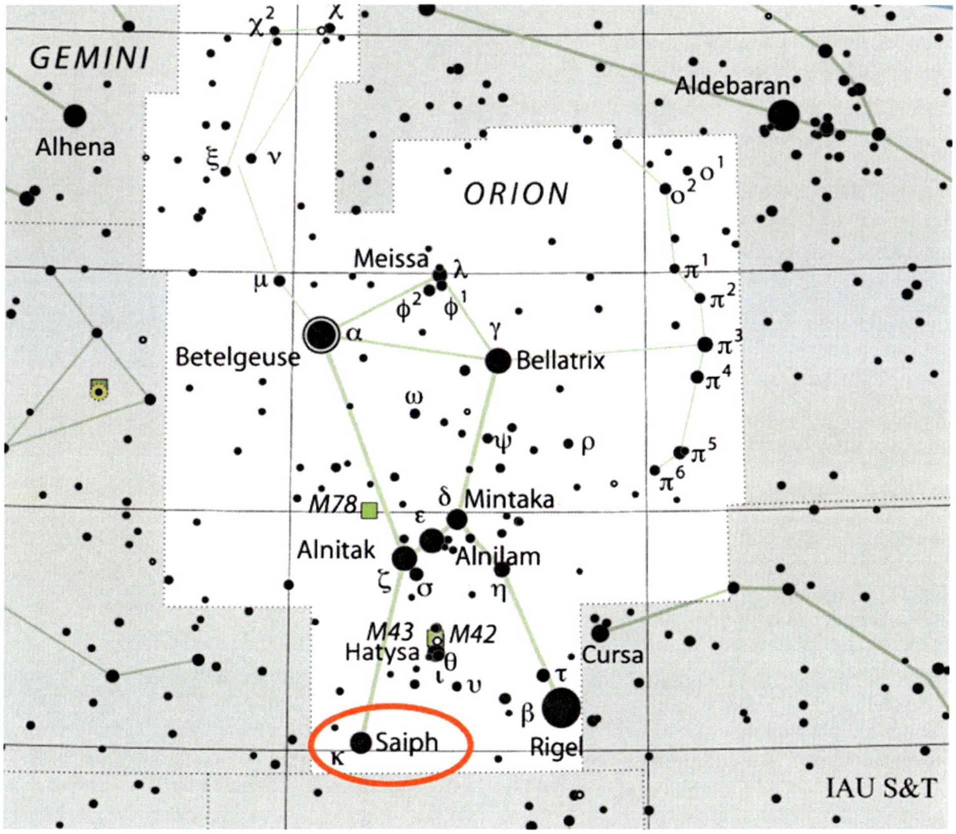

VISUAL ASPECTS:

Saiph is a supergiant whose luminosity is about 49,000 times greater than that of the Sun. This high intrinsic brightness enables us to see Saiph shine with an apparent magnitude of 2.1, even at the star's considerable distance of 650 light years. It is blue-white with a surface temperature of approximately 26,500 K.

PHYSICAL NATURE:

Saiph's mass is estimated at 15.5 times that of the Sun. Its diameter, 22 times wider than the Sun, indicates a volume 10,648 times the Sun's size. Saiph's spectral type is B0.5 and its luminosity classification is supergiant Ia.

Saiph marks the southeastern corner of a lobe in the Orion Molecular Cloud Complex. Studies of this star's radial velocity reveal Saiph is moving away from us at the rate of 21 kilometers per second, having been ejected in that direction during an early stage of evolution within the cloud complex.

Photometric classification of Saiph and a stellar host of others, combined with astrometry, radio, infrared, X-ray observations, integrated with marvelous theoretical conceptualizations have generated sophisticated knowledge about processes of stars and their evolution.

Orion Loop and Great Nebula M42 - Credit: Hewholooks

Astronomers and astrophysicists are our representatives as they explore the universe. Their imaginations, intellects, and instruments enable visits to the distant shores of wondrous worlds. We can join them, exercising our own imagination, knowledge, and devices, purposefully searching the nature and meaning of this grand endeavor.

SAIPH - ADDITIONAL DATA:

Catalogue Designations: Kappa Orionis, κ Ori, 53 Orionis, 141 G. Orionis, HD 38771, HIP 27366, HR 2004, SAO 132542
****

Parallax: 5 mas
Distance: 275 parsecs, 650 light years
****

Apparent magnitude: 2.1
Absolute magnitude: -4.4
Luminosity: 49,000 solar
Luminosity class: supergiant Ia
****

Apparent color blend: blue-white

Peak color: blue/blue-white
Color index (B-V): -0.18
Spectral type: B0.5
Surface temperature: 26,500 K
****

Diameter: 22 solar
Mass: 15.5 solar
Coordinates: (J2000.0) 05h48m -09°40' / 86.94 -9.67

Transit at 9 p.m. February 2
Proper motion: +1 -1 mas/year
Radial velocity: +21 km/sec.
****

Prequel – When We First Met

I had just returned from a mid-Atlantic solar eclipse cruise having a totality duration of four minutes and five seconds. My mind was on fire to learn astronomy – I sought more expertise from the Hayden Planetarium. AND THERE HE WAS! My teacher – with a golden tan obtained from playing tennis and volleyball with pals on Quogue beach, Long Island – fueling his play with a tuna salad sandwich each day, rather than being on the eclipse cruise.

How that happened was one of our first conversations! Steve was very collegial in helping me begin an astronomy club where I taught. Soon surfacing was our mutual love of childhood stargazing!

We, the city mouse observing celestial night from a rooftop in Brooklyn and the country mouse looking up from a spread thrown on the ground. Steve identified stellar objects by name having learned them from library books, astronomy magazines, and visits to the planetarium. I and my companions, Transco kids of Moore, South Carolina identified them as characters in a night magic carpet adventure. The meetup for going to the stars was ASAP after homework and duties, rush out and get aboard Mother's donated blanket. Positioned more in the center meant a human shield against blood thirsty mosquitoes. Off we went over the Milky Way. Sometimes the blanket was a magic carpet, sometimes a space ship, sometimes we travelled to say hello to the Moon. The navigator – storyteller for the night's trip counted us down ten, nine, eight, seven, six, five, four, three, two, one ... I didn't know the names of celestial entities, but I would one day! Little did I dream I was looking at my future life.

Twenty or so Transco kids moved around with rural joie de vivre, the extended family of Transcontinental Gas Pipeline Company. We jumped with towels and swimsuits into the company pickup after we persuaded a Dad to drive us for swimming in Silver Lake Park. We sang at the top of our lungs a medley of great tunes like "I've Been Working on the Railroad" all the way there and all the way back. Some of us rode bikes to visit the Kuhn's farm. They were retired New York restauranteurs who stabled horses that they were happy to have us exercise and tack. We learned to ride English and Western saddle, and raced bareback across the lower fields. While foraging wild blackberries for yummy Mom cobblers a merry band of us wandered off into the woods, discovering a little paradise. It was a cascade of water holes. The water ran over rocks trailing down one side that made a natural slide. There were vines to swing out on, release and drop into the deepest pool. What unspeakable fun! That is until parents found out, and said it must be fun swimming with water moccasins and all those other snakes. From then on, no more glorious water hole slides, we were back to the supreme paradise of night sky adventures.

Moore was barely a dot on the map between the city of Spartanburg and the town of Woodruff. Moore proper was a one stop all-purpose gas station, grocery, and post office. Timing was critical, otherwise you could be stopped by a passing train of many cars feeling that your youth was passing with it before its little red caboose appeared. You could then continue down route one a few miles before getting to Transco Natural Gas Compressor Station and the company houses of Superintendent Elmo Knudsen and five engineers who ran the plant 24/7. I fondly remember us, the Transco family, the Knudsens, Meyers, McCauleys, McGreeveys, Elgins, and the Adairs.

Once the Elgin girls, my superstar sister Eleanor three years ahead of me in Roebuck school, I, and our parents Melba and Frank were invited by Virginia and Elmo Knudsen to vacation in their Hilton Head Beach house Elmo had designed and built. One day the manager of the newly opened Hilton Head Hotel walked over to invite Eleanor and me to have lunch on their terrace. That would take second prize to an afternoon swim when whales passed by. From our vantage they looked like giant elephants breeching the water's surface. In those years, both there at the beach and back home, we were fortunate to have skies dark enough to see over a thousand stars on the clearest of nights.

On the other hand, Steve gazing from his rooftop under an urban light bubble could see at best only a few dozen naked eye stars. Therefore, Moon and planet spotting was the primary observing fare even after he got his first "real" telescope a 3.5-inch "Skyscope" reflector. The instrument first caught Steve's eye in a Hayden Planetarium display case and when his Dad went over to the manufacturer's 475 Fifth Avenue Manhattan address, he learned the place was just a mail drop, but got a phone number and discovered the telescope was made by two brothers named Herrick in their home basement workshop on Thirteenth Avenue, Brooklyn, just a mile and a half up the same street from the Beyer's apartment!

A meeting was arranged; a demonstration in the makers backyard provided a vivid view of the first quarter Moon; and a fine little telescope was taken home. In another bit of serendipity, Richard S. Luce, long time leader of Optical Division telescope makers at the Planetarium also happened to live on 73rd Street, just a few houses down from Steve's family. That eventually led to Steve making his own telescope mirror, working at the Planetarium, and meeting me!

Of those many nights observing on the roof with the telescope, often in early years with his parents, he remembers one particularly fine evening. On the trip home from work as a stationary engineer at the Kent Avenue power house that provided electricity to the New York City Subway system, his dad Svend noticed in the western sky a triple conjunction of Venus, Jupiter and the crescent Moon, and then alerted Steve so he too could experience that magnificent sight.

Otherwise, Steve remembers engaging in astronomical observations up on his roof with only three other people. They were his cousin Louis Beyer, (for whom telescopic lunar photography was demonstrated), a friend Stuart Hatfield (tracking a late evening total lunar eclipse), and NYU Prof. Karen Gillespie, his friend Dr. Robert Gillespie's Mom, (when one morning she too was spotting elusive Mercury during twilight).

Arcturus April 5

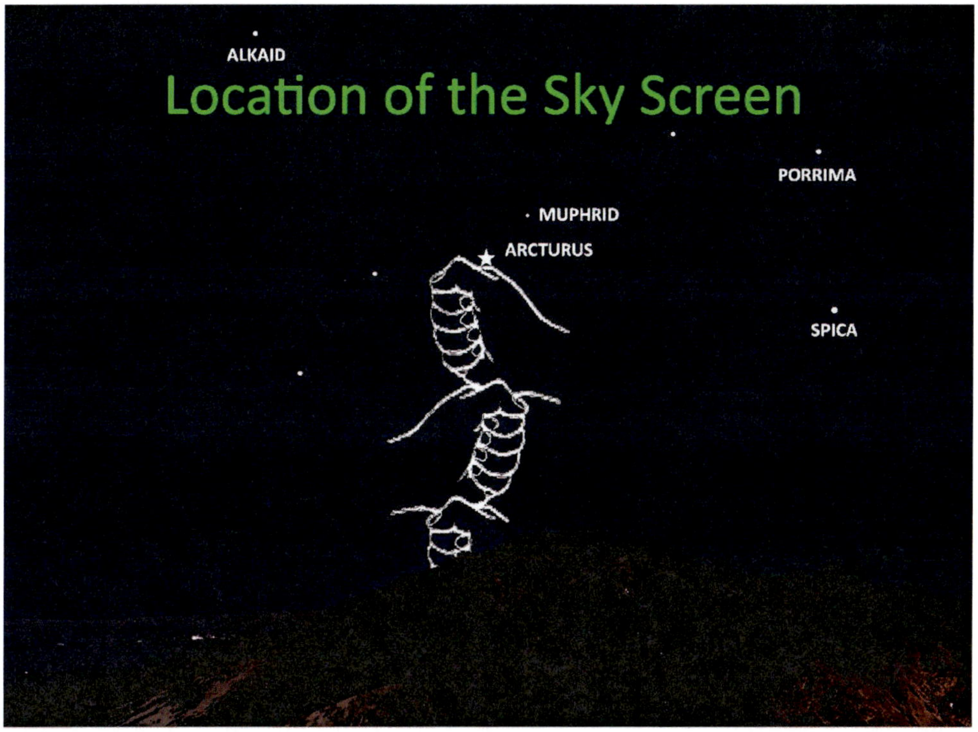

East-Northeast East East-Southeast

Bootes *constellation*
April-September *prime season*

East-Northeast East East-Southeast *–Sky Screen horizon vista range*
Arcturus – *Traditional Star Name (from the International Astronomical Union's 2018 list)*
arc-TO-russ –Prounciation (suggested)
-0.05 mag. – *Apparent Magnitude, star's brightness as seen from Earth*
yellow giant – *Apparent color blend and luminosity class*
37 ly – *Distance expressed in light years*
very high proper motion – *Special aspect of the featured star*
Catalogue Designations:
Alpha Bootes – *Bayer Designation, Greek letter and generic form of constellation name*
α Boo – *Bayer Abbreviation*
16 Bootes – *Flamsteed Catalogue number*
HD 124897 – *Henry Draper Memorial Catalogue*
HIP 69673 – *Hipparcos Guide Star Catalogue*
HR 5340 – *Harvard Revised – (Yale Bright Star Catalogue)*
SAO 100944 – *Smithsonian Astrophysical Observatory*
A405 – *The Star Guide date code*
****
Parallax: 89 mas – *distance in milliarcseconds parallax angle = 0.089 arcsecond*
Distance: 11 parsecs – *1/0.089 = 11.235955 ~ 11 parsecs*
37 light years – *11 x 3.26 = 35.87 ~ 37 light years*
****
Apparent magnitude: -0.05 – *directly measured by means of photometry*
Absolute magnitude: -0.32 – *calculated using known distance and apparent magnitude ('as if seen at 10 parsecs)*
Luminosity: 115 solar – *comparison (Sun = 1.0)*
Luminosity class: giant III – *this star's spectrum compared to MK "anchor point" stars*
****
Apparent color blend: yellow – *subjective examination of defocused star image*
Peak color: yellow-orange – *wavelength indicated by temperature, Wein's law*
Color index (B-V): 1.23 – *the star's magnitude observed and recorded through a B (blue) c. 445 nm wavelength filter, minus its magnitude measured through a V ("visual" = green-yellow) c. 551 nm filter*
Spectral type: K0 – *from spectrum comparison to MK standard stars*
Surface temperature: 4,200 K – *determination based on a star's color index, etc.*
****
Diameter: 25 solar – *25 times (the Sun's diameter of 868,925 miles), or 25 times (1,398,400 kilometers). The factor "25" may be obtained by applying the Stefan-Boltzmann law*
Mass: 1.5 solar – *1.5 times (Sun's mass: 5.97 x 10^{24} kilograms). The featured star's mass relative to the Sun. In the case of Arcturus "1.5", is estimated by applying Newton's version of Kepler's Third Law. That may be accomplished directly to a star if it's part of a binary system whose distance from Earth, and the component stars' angular separation, and orbital period*

have been measured. If that is not possible, a star's mass may be estimated by matching it with a reference binary star having similar spectral properties and directly measurable properties.
Coordinates: (J2000.0) 14h16m +19°11' / 213.92 +19.18 – ***east-west, north-south***
Transit at 10 p.m. June 11 – ***when a featured star is at the meridian – along a vertical line from horizon to zenith*** **(seen from the defining longitude of a time zone)**
Proper motion: -1093 -2000 mas/year – ***annual change of position in Right ascension, eastward (+), westward (-); and declination, northward (+), -southward (-). Units "mas" are thousandths of an arcsecond.***
Radial velocity: -5 km/sec. – ***rate of direct line travel toward (-) or away (+) from Earth. Measured by means of Doppler shifts of a star's spectral lines compared to laboratory wavelength measurements, where there is no relative motion between the light's source and observer.***

Greek alphabet

Αα	Alpha	Νν	Nu
Ββ	Beta	Ξξ	Xi
Γγ	Gamma	Οο	Omicron
Δδ	Delta	Ππ	Pi
Εε	Epsilon	Ρρ	Rho
Ζζ	Zeta	Σσς	Sigma
Ηη	Eta	Ττ	Tau
Θθ	Theta	Υυ	Upsilon
Ιι	Iota	Φφ	Phi
Κκ	Kappa	Χχ	Chi
Λλ	Lambda	Ψψ	Psi
Μμ	Mu	Ωω	Omega

Credit: Wikipedia

Appendix B Sky Screen Horizon Images

Credits: Adobe Stock Images - asi / Shutter Stock Images – ssi
Andrea & Steven Beyer – asb / Eric Elgin Beyer - eeb
Cover Images: NASA, ESA

Dates mo/da	Star Names	Scene Descriptions Credits
1002	Mirfak	Paris, Pont de Grenelle-Cadets de Saumur - asi
1004	Algol	western style windmill - asi
1017	Epsilon Persei	ranch house bright - asi
1021	Fomalhaut	hills and house lights - asi
1022	Diphda	Venice - asi
1024	Alcyone	mountains distant village - asi
1027	Menkar	cruise ship at dock -asi
1102	Capella	mountain towns - asi
1105	Hassaleh / Iota Aur	tree line scene - ssi
1110	Aldebaran	lower Manhattan - asi
1114	Menkalinan	Verrazzano-Narrows Bridge – eeb
1115	El Nath	Austin, Texas - asi
1119	Mahasim / Theta Aur	Big Bend National Park -asi
1201	Bellatrix	house in trees and snow - asi
1203	Tejat / Mu Gem	waves and clouds - asi
1207	Cursa	mountains with snow - asi
1208	Betelgeuse	small house hills - asi
1209	Mintaka	snowy peak to right - asi
1210	Alnilam	mountain crestline - ssi
1212	Alhena	San Diego - asi
1213	Alnitak	house - asi
1214	Rigel	mountains with snow - asi
1216	Hatysa / Iota Ori	Philadelphia - asi
1217	Castor	Temple of Castor & Pollux, Roman Forum - asi
1221	Pollux	houses and trees - asi
1224	Saiph	house and tree - asi

Additional IllustrationsPage Subject Link Credit
4 Alpha Persei Open Cluster: https://commons.wikimedia.org/wiki/File:Melotte_20.jpg Credit: Martin Gembec

8 Algol light curve from TESS https://commons.wikimedia.org/wiki/File:Algol_TESS_lightcurve.png Credit: Warrickball

10
https://commons.wikimedia.org/wiki/File:Hubble_Sees_the_Force_Awakening_in_a_Newborn_Star_(23807356641).jpg Credit: ESA/Hubble & NASA, D. Padgett (GSFC), T. Megeath (University of Toledo), and B. Reipurth (University of Hawaii)

10 Perseus molecular cloud system https://commons.wikimedia.org/wiki/File:PIA23405-PerseusMolecularCloud-SkyLocation.jpg Credit: NASA/JPL/12 Fomalhaut b https://commons.wikimedia.org/wiki/File:Fomalhaut_b_dust_cloud.png Credit: NASA, ESA, and A. Gáspár and G. Rieke (University of Arizona)

13 Johann Bayer, Uranometria 1603, Constellation Piscis Austrinus file:///C:/Users/froyc/OneDrive/Desktop/SGCS%208.511/Uranometria/Ioannis%20Bayeri%20Uranometria%20omnium%20asterismorum%20continens%20schemata,%20nova%20methodo%20delineata%20aereis%20laminis%20expressa.pdf Credit: Dr Rafael Ball, ETH-Bibliothek Zürich

17 Pleiades Open Star Cluster in Taurus-
https://upload.wikimedia.org/wikipedia/commons/thumb/4/4e/Pleiades_large.jpg/2560px-Pleiades_large.jpg Credit: NASA, ESA, AURA/Caltech, Palomar Observatory

18 Brown Dwarf Comparisons https://commons.wikimedia.org/wiki/File:BrownDwarfs_Comparison_01.png Credit: MPIA/V. Joergens

20 T Tauri example – XZ Tauri binary eruption https://commons.wikimedia.org/wiki/File:Hs-2000-32-a-animated.gif Credit: John Krist (STScI), Karl Stapelfeldt (NASA Jet Propulsion Laboratory), Jeff Hester (Arizona State University), Chris Burrows (ESA/STScI)

25 Hyades Open Star Cluster in Taurus https://commons.wikimedia.org/wiki/File:Hyades_40%C2%B0N.png Credit: Roberto Mura

28 Epsilon Aurigae light curve https://commons.wikimedia.org/wiki/File:Epsilon_Aurigae_eclipse_light_curve.png Credit: Lithopsian

28 Johann Bayer, Uranometria 1603, Constellation Auriga Credit: Dr Rafael Ball, ETH-Bibliothek Zürich

30 Iota Aur Hass MKK Atlas anchor point star https://ned.ipac.caltech.edu/level5/ASS_Atlas/frames.html Credit: Caltech

36 Harvard "Women Computers" c. 1879 https://commons.wikimedia.org/wiki/File:Astronomer_Edward_Charles_Pickering%27s_Harvard_computers.jpg Credit: Harvard College Observatory

36 Menkalinan light curve: https://commons.wikimedia.org/wiki/File:BetaAurLightCurve.png Credit: PopePompus

45 Hertzsprung-Russell Diagram example: https://commons.wikimedia.org/wiki/File:HR-diag-instability-strip.svg Credit: Persus

47 Orion molecular clouds https://en.wikipedia.org/wiki/File:Orion_Head_to_Toe.jpg Credit: Rogelio Bernal Andreo

52 Johann Bayer, Uranometria 1603, Constellation Eridanus https://huntington.org/educators/learning-resources/spotlight/uranometria Credit: The Huntington Library

56 Bayer Uranometria – Constellation Orion Credit: The Huntington Library

64 NGC2264 https://commons.wikimedia.org/wiki/File:NGC_2264_by_ESO.jpg Credit: ESO

74 BN Object in Orion https://commons.wikimedia.org/wiki/File:ALMA_views_a_stellar_explosion_in_Orion.jpg Credit: European Southern Observatory

82 Barnard Loop – Emission Nebula in Orion https://commons.wikimedia.org/wiki/File:BarnardLoopHunterWilson.jpg Credit: Hewholooks

Glossary

Aberration of Starlight. A slight systematic annual variation in a star's apparent position, caused by Earth's orbital velocity relative to the speed of light.

Absolute magnitude. The apparent magnitude a star would have if it were seen at a standard distance of 10 parsecs (32.6 light years).

Absolute zero. The zero point of the Kelvin temperature scale, equal to -273.16° Celsius or -459.69° Fahrenheit.

Absorption. The manner in which matter absorbs electromagnetic energy.

Absorption spectrum. Characteristic dark lines, bands, or graphic valleys displayed on a continuous spectrum of colors when light is passed through a gas.

Acceleration. A change of speed and/or direction of motion.

Accelerating universe. Situation by which dark energy is accelerating its overall expansion.

Accretion disk. A disk surrounding a star or other celestial object, formed by orbiting particles or gas attracted to the object's vicinity by its gravity.

Accreting Neutron Star. A neutron star that is attracting and absorbing mass from its environment.

Active galactic nuclei. The very bright centers of some unusual galaxies; suspected of being caused by violent accretion at central supermassive black holes; quasars are the most extreme examples.

Adaptive Optics. Mechanically manipulated telescope optics; thereby compensating for image degradation caused by "twinkling" of celestial objects.

Altitude. Angular separation between the horizon and a celestial object.

Angular diameter. The apparent diameter of a celestial object, expressed in degrees, minutes, or seconds of arc (see Degree).

Angular momentum. Momentum of spin; calculated as the product of a turning object's mass times rate of spin, times its average radius.

Angular resolution. Ability of a telescope to discern small apparent separations.

Angular separation. The apparent distance between two stars or other celestial objects, expressed in angular measurements.

Angular size. The apparent size of a celestial object expressed in angular measure.

Aperture. The usable diameter of a telescope's objective lens or mirror.

Apparent brightness. A measure of the amount of light per unit surface area reaching an observer

Apparent magnitude. The brightness of a celestial object expressed in a system in which a difference of one magnitude represents a brightness factor of about 2.5.

Arcsecond. A unit of angular measurement equal to 1/3,600 of a degree of arc 1/60 of an arcminute, and 1,000 milliarcseconds.

Asterism. An eye-catching formation of stars, which may (or may not) belong to a single constellation or might be in several adjacent constellations.

Astrometry. A traditional branch of astronomy that concentrates on measurements of the positions and motions of celestial objects. It is sometimes known as spherical or positional astronomy.

Astrophysics. The branch of astronomy that concentrates on the physical nature of celestial objects.

Atmosphere. An envelope of gas that surrounds a celestial object.

Autumnal Equinox. An invisible point on the celestial sphere, located in the constellation Virgo, defined by the intersection of the ecliptic and the celestial equator. The sun's arrival at this point, on or about September 23, marks the beginning of autumn in the Northern Hemisphere.

Azimuth. Positions along the ecliptic indicated in units of angular degrees and their subdivisions; measured clockwise from the north point (90°, 180°, etc.).

Barred spiral galaxies. Galaxies such as the Milky Way whose spiral arms curl out from two protrusions extending from their central bulge.

Baryonic matter. Ordinary matter such as what comprises us and everything we can see.

Big Bang. The event believed to have originated our universe.

Binary star. A double star whose components are gravitationally bound to each other and revolve around a common center of mass.

Black Body Radiation. Electromagnetic radiation having a certain wavelength, determined by the temperature of an opaque object that is its source, such as the filament in a light bulb.

Black hole. An object whose mass has become so highly concentrated that it theoretically occupies no space at all. The gravitational field surrounding a black hole is believed to be so intense that even light may not escape from a region immediately surrounding the concentrated mass.

Blink comparator. A device used for comparing images of the sky, with the intention of quickly revealing presence of relatively fast-moving objects such as asteroids, comets, or other Solar System objects traversing fields of far more distant stars.
BL Lac objects. A type of active galaxy whose radio jets happen to currently be aimed at Earth.
Blue shift. A Doppler shift of spectral lines towards the shorter-wavelength, blue end of the spectrum; caused by the light source's motion towards the earth.
Bok globule. An approximately spherical zone of relatively concentrated dust and gas that may eventually in fall to become a new star.
Brown Dwarf. An object with less than 0.08 solar mass, too small to have sufficient gravity to crush its center sufficiently to initiate nuclear fusion – such fusion being the defining attribute of stars.
Bulge of a galaxy. Spheroid assembly of old stars in the central zone of a spiral galaxy.

Carbon Stars. These are stars close to the end of their existence, where carbon generated by thermonuclear fusion has significantly accumulated in their atmospheres.
Carbon Nitrogen Oxygen cycle (CNO). Thermonuclear reactions in high and intermediate mass stars involving the fusion of hydrogen into helium plus energy release.
Celestial coordinates. Sky position indicating system, based on the celestial equator and poles, and orthogonal units of right ascension and declination.
Celestial equator. The great circle on the celestial sphere that lies halfway between, 90 degrees of arc from each celestial pole.
Celestial mechanics. The branch of astronomy most concerned with the interactions of forces and motions of celestial objects.
Celestial meridian. The great circle on the celestial sphere passing through celestial poles, as well as the zenith and north and south points on the horizon of an observer's sky. A celestial object's transit of the celestial meridian represents its highest sky elevation.
Celestial poles. The two imaginary points on the celestial sphere defined by intersections with extensions of the earth's axis of rotation.
Celestial sphere. An apparent sphere of indefinite size upon which celestial objects appear to be located.
Center of mass (gravity). The point at which all the mass of an object or system of objects may be considered to be concentrated.
Central galaxy. A particularly massive galaxy at the center of a galaxy cluster such as the elliptical M87 in the heart of the Virgo cluster, evidently enlarged by absorption of other galaxies.
Cepheid variable. A type of highly luminous, pulsating supergiant stars, named after their prototype, Delta Cephei. The periods of light variations in these stars are directly related to their intrinsic brightness, and as a result Cepheids are used to estimate distances up to about 10 million light years from Earth. They are also known as classical or type I Cepheids.
Chandrasekhar limit. An upper limit, of about1.4 solar masses, on the theoretical mass of a white dwarf star. If a dying star is more massive it violently collapses, explode as a supernova with total devastation or perhaps become a neutron star or black hole.
Charge-coupled device (CCD). An electronic accessory used to increase the sensitivity of a telescope in recording faint visible light and other forms of electromagnetic radiation.
Chromosphere. The layer of a star above its photosphere and beneath the corona.
Circumpolar stars. Stars in the vicinity of the celestial pole that are always above the horizon for observers at a particular latitude.
Circumstellar shell. A shell of dust or gas that surrounds a star.
Cluster of galaxies. A group of up to several thousand galaxies, bound together by their mutual gravitational attraction.
Colure. One of two great circles on the celestial sphere that pass through the poles as well as points on the ecliptic that mark the sun's position at the start of each season. The equinoctial colure intersects both the vernal and Autumnal Equinoxes. The solstitial colure passes through both the summer and winter solstices on the ecliptic.
Condensation. consolidation of matter into a smaller volume.
Configuration. A grouping seen in the sky formed by the temporary juxtaposition of two or more planets, stars, and/or the moon.
Constellation. One of eighty-eight areas in the sky, whose boundaries, defined in 1930, specify regions on the celestial sphere.
Continuous spectrum. A continuous array of colors or other wavelengths of electromagnetic radiation. Continuous spectra are produced by hot solids, liquids, or highly compressed, opaque gases such as that which constitutes stellar interiors.
Core (of a star). The central region of a star. In main sequence stars, all energy is produced via fusion of hydrogen into helium in their cores, where temperatures are at least 15 million Kelvins.

Cosmic microwave background (CMB). This microwave eneGlossaryrgy reaches us from all directions in the universe. It was discovered during 1964 by Arno Penzias and Robert Wilson. It consists of stretched out photons of light from an early stage in the formation of the universe. The CMB dates from the epoch of recombination when electrons and protons formed hydrogen atoms about 378,000 years after the Big Bang. Thereby, space was essentially cleared of a fog of freely separated electrons and protons, thereby enabling photons to travel in direct lines without frequently encountering and being scattered by sub-atomic particles. The CMB corresponds to thermal (black body) radiation peaking at 2.73 K.
Cosmic rays. High energy protons and atomic nuclei that reach earth from space, traveling at speeds approaching that of light. Supernovae explosions are their primary source.
Cosmological Principle. The concept that on the largest scales, matter is distributed homogeneously through the entire universe, which has no center and no edge.
Cosmology. Study of the origin, structure, and evolutionary processes of the universe.

Dark energy. The energy or other "stuff" that causes the universe's expansion to accelerate.
Dark matter. Matter currently revealed only by means of its gravitational influences on ordinary (baryonic) matter,
Dark nebulae. Relatively dense cold regions of interstellar dust that obscure the light of stars within the nebulae or behind them.
Declination. The coordinate on the celestial sphere which measures angular distances north or south of the celestial equator.
Degenerate object. A celestial object such as a neutron star, or white dwarf, where neutrons in the first example and electrons in second type, are so closely packed they resist gravity's tendency to drive them closer together. Degeneracy rather than heat resists further collapse of those object.
Degree (angular). 1/360 of the circumference of a circle. A degree of arc on the celestial sphere is equal to about twice the apparent diameter of the Full Moon.
Density. A measure of the concentration of matter, expressed by unit mass per unit volume.
Density-wave theory. A concept used to explain the spiral structure of galaxies.
Diffuse nebulae. An irregular cloud of gas or dust in a spiral galaxy such as the Milky Way.
Disk (of a celestial object). The apparently circular cross-section of a spherical celestial object.
Disk (of a galaxy). The flattened, circular region containing the majority of stars, interstellar dust and gas, as well as star forming parts of a spiral galaxy. It contains the category of stars known as population I.
Doppler Effect (shift). Variation in wavelength and frequency of an electromagnetic wave such as visible light or ultraviolet caused by relative motion between source and observer.
Double shell fusion star. In this type of star an inert carbon core is covered by a shell that is fusing helium into carbon, which in turn is inside a hydrogen to helium burning shell.
Dwarf galaxy. A galaxy containing less than about ten billion stars.

Eclipse. The complete or partial disappearance of a celestial object behind another or the appearance of the shadow of the Earth on the Moon.
Eclipsing binary star. A star, such as Algol in a multiple star system, periodically eclipsed by one or more companions seen along our line of sight,
Ecliptic. The apparent path of the sun across the celestial sphere. .
Electromagnetic radiation. Radiation produced as a result of variations in electric and magnetic fields. The electromagnetic spectrum includes gamma rays, X-rays, ultraviolet light, visible light, infrared light, and radio energy. All forms of electromagnetic radiation have a speed in a vacuum of about 3×10^5 kilometers per second, approximately 186,000 miles per second.
Electromagnetic spectrum. The range of wavelengths manifested by electromagnetic energy.
Electron. A subatomic particle with a negative electric charge and which normally moves in zones surrounding the nucleus of an atom.
Element (chemical). A substance that cannot be broken down into its components through chemical actions.
Elephant-trunk structures. Long, dark cones that may be associated with star formation processes. They are seen in some nebulae where H II regions of heated gas are expanding and flowing past comparatively dense regions of cool dust and gas. When seen from the end, elephant-trunk structures may appear as dark spots called globules.
Ellipse. A closed curve representing the orbital paths of planets, satellites, and components of multiple star systems.
Elliptical galaxy. A type of galaxy having an elliptical cross-section containing relatively small amounts of interstellar gas and dust.
Emission line. A bright spectral line produced at a specific wavelength of electromagnetic radiation.
Emission nebula. A cloud of gas having temperatures of about 10,000 K, with a spectrum characterized by emission lines.
Emission spectrum. A spectrum that consists of emission lines produced by hot glowing gas of low density.

*Energy. A*bility to do work, to move things.

Epoch. A specific date and time used as base for calculating revised numbers for astronomical quantities that systematically change over time such as a star's celestial coordinates changing due to proper motion, or a planet's orbital elements changing because of gravitational influences of other planets. The current standard is J2000.0, January 1, 2000 at noon TT (Terrestrial Time).

Equatorial mounting. A telescope mounting having one axis parallel to the Earth's axis so that compensation may be more easily made for the Earth's rotation.

Equinoctial colure. The meridian circle on the celestial sphere which passes through the celestial poles and the Sun's location on the dates of the equinoxes.

Equinox. Specific date and time when the Sun's position in the sky relative to an encompassing celestial reference frame is indicated. The current standard is J2000.0, January 1, 2000 at noon TT (Terrestrial Time). The "J" indicates Julian years consisting of 365.25 days are being used.

Era of Planck. The earliest time in the universe's existence, prior to 10^{-43} seconds immediately following the Big Bang. During this brief episode basic field forces of nature, gravity, strong nuclear, weak nuclear, and electromagnetic force were entwined in a single super-force.

Era of GUT (Grand Unified Theory) According to theory, this was an extremely early period in the universe's existence, from 10^{-43} second until 10^{-38} second after the Big Bang. During this episode gravity emerged as a separate force, leaving the strong nuclear, weak nuclear force, and electromagnetism together. At the end of the GUT era, the universe had cooled sufficiently so that the strong nuclear force was able to start separating and begin emerging as a separate and distinct force. The electromagnetic and weak nuclear force remained in what is known as the electroweak force. Separation of the strong nuclear force released vast energy which caused the nascent universe to swell at a rate faster than the speed of light in the process called cosmic inflation.

Event horizon. The "surface" surrounding a black hole at which the escape velocity is equal to the speed of light. Its distance from the black hole's center is about three kilometers for every solar mass in the object.

Exoplanet (Extrasolar planet). A planet beyond the Solar System

Extragalactic. Outside the Milky Way Galaxy.

Eyepiece. The lens or set of lenses in a telescope where you place your eye to look through the instrument.

Fermi's Paradox. If extra-terrestrial intelligent beings exist, why haven't we seen any?

First Point of Aries. A traditional name for the Vernal Equinox, the Sun's location on the celestial sphere at the start of the spring season in the Northern Hemisphere. This term originated when the Vernal Equinox was actually located in the constellation Aries; precession has now shifted this equinox to the constellation Pisces, the Fish.

Flare star. A low mass spectral type M star demonstrating very strong surface flares.

Force. A push or pull causing a change in an object's speed and/or direction of travel. Defined by Newton as mass times acceleration (F=ma).

Full Moon. The lunar phase at the point of its orbit at which it is 180 degrees of arc from the Sun. Then its disk appears fully illuminated when viewed from Earth.

Fundamental forces. Gravity, electromagnetic, strong nuclear, and weak nuclear forces

Fusion (Thermonuclear). The joining together of lighter atomic nuclei to form heavier ones, accompanied by a transformation of matter into energy. This is the basic means of energy production in stars.

Galactic equator. A great circle on the celestial sphere representing the mid-line of the Milky Way.

Galactic Halo. Spherical region encompassing Milky Way's disk. The Halo comprises stars including members of globular star clusters, which are of very early Galactic generations, therefore they possessing relatively low levels of elements other than hydrogen and helium. Halo stars are designated as being Population II objects.

Galactic poles. Two points on the celestial sphere, each located 90 degrees of arc from the Galactic equator.

Galactic star cluster. Another name for an open star cluster, which may contain up to several hundred stars in irregularly shaped groups located in the disk of the Milky Way.

galaxy. A gravitationally bound assembly of between several million to many hundreds of billions, even trillions of stars.

Galaxy. The name of the galaxy that contains the Solar System; also known as the Milky Way Galaxy. By convention "our" Galaxy is always capitalized.

Gamma-ray burst. A violent blast of gamma rays originating in deep space.

Giant molecular cloud. A huge mass of cold interstellar gas containing up to a million solar masses of matter. They contain the raw material of stars and organic molecules.

Giant star. A star in late stages of its evolution, having an inflated diameter and high luminosity, in large part due to its enhanced surface area.

Globular cluster. A gravitationally bound, spherical group of hundreds of thousands of stars, which orbits the center of a galaxy in an encompassing volume of space known as a galactic halo. Stars in these clusters are believed to have been formed in the earliest days of a galaxy's existence.

Globule. A small, dense concentration of dust and gas, possibly a site where new stars may form.
Goldilocks Zone. A circumstellar region where temperatures allow liquid water to exist on the surface of an orbiting planet having sufficient atmospheric pressure. Also known as the habitability zone.
Gravitation. A property by which matter is attracted to other matter.
Gravitational collapse. An unchecked condensation of matter produced by the influence of gravity.
Gravitational energy. Energy released when matter in a system becomes more concentrated.
Gravitational lens. A phenomenon in which the gravitational field of a celestial object such as a galaxy acts as a lens that focuses the image of a more distant entity viewed along nearly the same line of sight.
Gravitational Waves. Traveling through space at the speed of light, these subtle forms of energy are generated by all moving objects. As yet they have only been detected as probable consequence of the cataclysmic collision of distant black holes.
Great circle. A circle on a spherical surface in which the plane of the circle passes through the center of the sphere.
Gyrochronology. Use of a type F, G, K, or M main sequence star's rate of rotation (V sin i), relative to that of the Sun, to estimate approximate its age.

H I region. An interstellar cloud of cool hydrogen gas at a temperature of less than about 100 K.
H II region. An expanding region of interstellar hydrogen surrounding hot, young stars heated to temperatures of about 10,000 K by ultraviolet radiation from them. H II regions are also known as Stromgren spheres.
Habitability zone. The region surrounding a star where surface temperature of a planet could permit the existence of water in liquid form, and where life as we know it might be able to exist.
Halo (of a galaxy). A spherical region surrounding the disk of a spiral galaxy. The halo is thinly populated by stars that orbit the galactic center either singly or as members of globular clusters.
Hawking radiation. Radiation hypothesized to be produced by the evaporation of a black hole.
Hayashi lines. Theoretical paths of protostars on the H-R diagram as they approach the main sequence.
Heavy elements. In general astronomical usage, chemical elements other than hydrogen and helium.
Helium flash. When the inert helium core of a low mass star between 0.8 and 2.0 solar masses abruptly begins runaway fusion activity, converting helium into carbon and energy via the triple alpha process.
Herbig-Haro objects. Patches of nebulosity surrounding nascent stars, which are formed as bi-polar jets of material ejected by the protostars collide with interstellar material near the surface of a molecular cloud.
Hertzsprung-Russell (H-R) diagrams. Charts illustrating correlations between the brightness and colors/surface temperatures of individuals in particular groups of stars.
High mass star. Those with an initial mass more than eight times that of the Sun and which are apparently destined to eventually explode as a supernova.
High-velocity star. A star whose velocity differs substantially from that of the Sun. These stars are generally temporary visitors to the Sun's vicinity in the galactic disk and are normally located in the halo region of the galaxy.
Horizon. The astronomical horizon is defined as a great circle on the celestial sphere, located 90 degrees from the zenith.
Horizontal branch. The zone of stars on an H-R diagram representing stars that are in the process of fusing helium.
Hour circle. One of the great circles on the celestial sphere that pass through the celestial poles and that are used to mark hours of right ascension in the equatorial coordinate system.
Hubble constant. A number used to correlate the redshift velocity of celestial objects with their distances from earth. Recent estimates of the Hubble constant range from the high 60's to low 70's in units of kilometers per second per million parsecs.
Hubble law. A statement of the relationship between velocity of separation from us and distances from Earth to remote galaxies. H_o (Hubble's constant) = velocity (kilometers per second)/distance (mega parsecs)
Hubble Space Telescope. The revolutionary Earth orbiting reflecting telescope launched in 1990 that vastly enhanced the quality of celestial observations at visual and near infrared wavelengths during its decades of service.
Hubble time. A period of 13.8 billion years, which represents the time required for the universe to have expanded from the primeval fireball to reach its present size.
Infrared radiation. A form of electromagnetic radiation with wavelengths in the range between those of red visible light and microwaves. Infrared radiation is also known as radiant heat.
Interferometer. A device designed to blend two or more beams of electromagnetic radiation from a single source in order to provide increased resolution and more information about the object's angular size.
Intermediate mass star. Objects with masses from about 2.0 solar to 8.0 solar.
These stars end their existence as a white dwarf within a relatively fast dissipating planetary nebula.
International Celestial Reference System (ICRS). Methodology defining a fundamental celestial reference system.
International Celestial Reference Frame2 (ICRS2). Used to specify positions of celestial objects. Centered at the Solar System's center of mass, its matrix is anchored by 295 defining sources, mostly quasars, located by very long baseline radio interferometry (VLBI).

Interstellar dust. Grains of matter of about 0.0002 millimeter in diameter (about the size of chalk dust or very fine clay) that obscure the light from background stars.

Interstellar gas. Gas of various kinds, mostly hydrogen, found in regions between stars.

Interstellar medium. Gas, dust, magnetic fields, and cosmic rays found between stars.

Ionization. The process in which one or more electrons are removed from the normal number in a previously electrically neutral atom.

Irregular galaxy. A galaxy that appears asymmetrical, lacking spiral, barred spiral, lenticular, or elliptical structure.

Irregular variable. A variable star with uncertain periods of light variation.

Island Universes. A transitional term used to describe galaxies beyond the Milky Way Galaxy. The expression was applied during early decades of the 20th century, as evidence mounted to support the notion a many other galaxies exist in a vastly larger cosmic realm, in which the Milky Way is not the entirety of the Universe, as had generally been believed before the work of Edwin Hubble.

Isochrone (clock) technique. Determination of star age in a particular cluster, based on aspects of relative stellar distributions seen on a Hertzsprung-Russell diagram

James Webb Space Telescope. Launched in 2021, featuring the largest space telescope optical ststem, especially tuned to infrared wavelengths enabling enhanced views through dust clouds and imaging of extremely redshifted early generations of galaxies.

Kelvin temperature scale. A temperature scale system whose units are of the same size as Celsius degrees, and whose zero point is equal to minus 273.16° Celsius and minus 459.69° Fahrenheit.

Kepler's laws. Three laws devised by Johannes Kepler in the early seventeenth century to describe orbital motions of planets.

Kinetic energy. Energy of motion.

Light (visible). Wavelengths between about 4000 Ångstroms (400 nm) and 7000 Å (700 nm) of electromagnetic radiation, visible to the human eye.

Light curve. Characteristic changes in the brightness of a variable star over a period of time.

Light-year. The distance that light travels during one year through a vacuum. A light-year is equal to about 9.5 trillion kilometers, 6 trillion miles and 63,000 A.U. (astronomical units).

Limiting magnitude. The magnitude of the faintest object visible to the unaided eye at a particular time, or through a certain telescope under ideal seeing conditions.

Local Group of galaxies. A group of about fifty gravitationally bound galaxies, including the Milky Way, the Great Galaxy in Andromeda, M33 in the constellation Triangulum, and many smaller members.

Local Supercluster. A vast collection of many clusters of galaxies, including the Local Group and the Virgo Cluster.

Lookback time. Length of time since light recorded on a telescopic image left its source. For example, how long it has been since a galaxy, appearing to us as a faint smudge on the Hubble Deep Field, would have been seen big, bright and with much detail by an observer near it, less than a million years after the galaxy emitted that light 12.8 billion years ago. The galaxy's tiny HDF image is a snippet of a portrait painted long ago that began a journey toward Earth 12.8 billion years.

Low mass star. Objects with less than about two solar masses. They wind up as white dwarfs inside the quickly dissipating shell of a planetary nebula.

Lucky imaging. Recording a lengthy sequence of images, and then selecting only those having the highest resolutions for combining in stacks, thereby minimizing individual random atmospheric distortions.

Luminosity. Intrinsic brightness of a star compared to that of the Sun.

Luminosity class. A category used to describe a star's luminosity and relative position on the H-R diagram.

Main sequence. The region of an H-R diagram that represents luminosity and spectral type of about 90 percent of the stars in the neighborhood of the Sun. During its main sequence stage, a star derives its energy from the thermonuclear fusion of hydrogen to helium in its core.

Main sequence turn-off point. This location along the sequence where a star leaves the line, moving to the final stages of existence with dramatic changes to its mode of energy production, as well as luminosity and surface temperature. The turn-off point for stars in a cluster is a key indicator the cluster's age.

Mass. A measure of the amount of material in an object or defined space.

Mass ejection. Process by which matter is ejected into space from the surface of a star.

Mass exchange. Gravitational transfer of mass from one star to a neighbor in a very close binary system,

Mass-luminosity relation. A graph that relates mass and luminosity of main sequence stars.

Meridian (celestial). A great circle on the celestial sphere that passes through an observer's zenith and north and south horizon points. The time of meridian transit represents a celestial object's highest elevation in the sky.

Messier catalogue. A listing of star clusters and nebulae prepared by Charles Messier during the late eighteenth century. These objects are often known by their "M" numbers.

Meteor. Often called a "shooting star," this phenomenon is an incandescent trail of gas emitted when a piece of rock or iron is heated by friction as it plunges through the earth's atmosphere.
Meteor shower. An encounter between the earth and a swarm of meteoroids that results in many meteors appearing to radiate from near one point in the sky.
Meteorite. A piece of rock or stone that survives its plunge through the atmosphere as a meteor and reaches the earth's surface.
Meteoroid. A small piece of rock or stone in orbit around the Sun.
Microwave. A form of very short-wave radio energy, with photons sized from millimeters to micrometers, located on the electromagnetic spectrum just beyond the infrared.
Milky Way. The faintly glowing band of light that circles the sky and represents our view of stars and nebulae in the disk of our galaxy. "Milky Way" is also used as the proper name of this galaxy.
Minute of arc. A minute of arc is equal to 1/60 of an angular degree or to 60 arcseconds.
Mira-type variable. A type of long-period, irregular variable stars, named after their prototype, Mira, the star also known as Omicron Ceti.
Molecular Bands. Tightly bunched set of spectral lines produced by molecules in a star's upper atmosphere.
Molecular cloud. A cool cloud in or near a galactic disk consisting mainly of molecular hydrogen at temperatures of about 10 to 50 Kelvins. The molecular clouds also contain traces of other gases, such as carbon monoxide, plus quantities of interstellar dust. These clouds are often star formation sites.
Molecule. The smallest particle of a chemical compound that retains the properties of the compound.

Nebula. A cloud of gas or dust in interstellar space.
Neutron star. A supernova remnant of extremely high density comprised almost entirely of neutrons. Some neutron stars manifest themselves as pulsars.
New General Catalogue (NGC). A catalogue of star clusters, nebulae, and galaxies.
Newton's laws. Fundamental laws of gravitation and mechanics devised by Isaac Newton in the late seventeenth century.
Nonbaryonic matter. Matter that is not the normal content of atoms.
Nova. A star that suddenly increases up to several hundreds of thousands of times in brightness. Novae may result when hydrogen streaming from the giant component of a close binary system accumulates on its white dwarf companion in sufficient amounts to trigger a thermonuclear explosion. A great burst of heat and light energy is unleashed, and the acquired mass is blown away from the surface of the white dwarf.
Nuclear fusion. The merging of two or more atomic nuclei to form a more massive nucleus and the conversion of some mass into energy.
Nucleosynthesis. The processes within stars whereby atomic nuclei are assembled from lighter nuclear components as a result of thermonuclear fusion.

OB association. A group of up to several hundred brilliant, young stars formed near each other in spiral arms of the Milky Way and other galaxies. An association contains too few stars to be gravitationally stable over the long term and the group eventually disperses after a few million years.
Objective lenses. The lenses of refracting telescopes that collect light and form images of objects that may be magnified by eyepiece lenses.
*Obscuration (interstellar). D*imming and reddening of light from celestial objects resulting from the intervention of interstellar dust.
Occultation. An eclipse of one celestial object by another.
Occam's razor. Notion that it is better to favor the simpler of two equally valid explanations of events.
Olbers' paradox. The concept that, if the universe is infinite and filled with stars, the sky should not be dark at night. In fact, it should be entirely as bright as the surface of the Sun!
Open (star) cluster (galactic cluster). A group of up to several hundred stars found in the disk and particularly in the spiral arms of the galaxy. Open star clusters are gravitationally bound and their member stars remain together in irregularly shaped groups.
Optical double. These are double stars that only appear to be close to each other but are not parts of the same, gravitationally bound system. Optical doubles are seen when stars that are at different distances from earth are viewed along nearly the same line of sight.

Parallax (stellar). A slight variation in a star's apparent position of a star on the celestial sphere, caused by a change in an observer's vantage point as earth moves to a new location along its orbit. Measurements of stellar parallax enable astronomers to accurately determine the distances of stars that lie less than about 500 light years from earth. Data from the new Gaia mission should extend that distance considerably. The trigonometric parallax of a star is equal to the angular distance between the earth and the sun, as observed from the star's distance. All stars beyond the sun have trigonometric parallaxes of less than one arcsecond. It is the fundamental method for determining distances of stars.

Parsec. The unit of space distances often used by astronomers. A parsec is defined as the distance from earth at which a star would have a parallax of one arcsecond. A parsec is equal to about 3.26 light years.

Period. The interval of time between successive occurrences oaf particular event.

Period-luminosity relation. The relationship between luminosities (intrinsic brightness, absolute magnitude) and duration of periods of light variation in variable stars. Cepheid stars for example with longer periods have higher luminosities than Cepheids with shorter periods. The period-luminosity relation is an important tool used to estimate distances in space.

Photometry. An astronomical method for measuring the intensities of various wavelengths of the electromagnetic spectrum of celestial objects.

Photon. An individual unit of electromagnetic energy, to which may be ascribed a particular frequency and wavelength.

Photosphere. The visible surface of a star.

Planet. A spherical celestial object that orbits a star and gravitationally dominates its orbital zone.

Planetary nebula. A sphere of luminous gas surrounding certain low and medium mass stars at the end of their evolution. These nebulae superficially resemble planets but actually are cross-sectional views of expanding gas shells ejected from energy depleted stars as they transform into white dwarfs.

Polar axis. Earth's axis of rotation or the axis of a telescope that is aligned with that of the earth.

Population I stars. Relatively young stars located in a galaxy's disk and especially in spiral arms. The Sun, with an age of about 4.6 billion years, is one of the older population I stars. These stars have been enriched by quantities of metals and other heavy elements formed during episodes of nucleosynthesis that occurred during the evolution of previous generations of stars in the galaxy's disk. The substance and activities of life on the earth are made possible due to the presence of heavy elements such as carbon, oxygen, and iron.

Population II stars. These stars are usually found in a galaxy's halo or in its central bulge. Occasionally the orbit of a halo star carries it through the galaxy's disk, as is the case with the star Arcturus. Population II stars are survivors from the early stages in the formation of galaxies, and as a result, they never had an opportunity to be enriched by significant quantities of heavy elements spewed into space by aging stars and incorporated into new stellar generations within a galaxy's disk.

Position angle. The angular distance measured from north through east on the celestial sphere that defines the position of a secondary star with respect to the primary in a double star system. '

Precession. The conical wobble of earth's axis of rotation over a period of about 26,000 years, caused by gravitational influences on earth from the Sun, Moon, and other planets of the Solar System. One effect of precession is a slow change in the positions of stars relative to earth-oriented coordinate systems.

Primeval fireball. The event that formed our expanding universe.

Proper motion. The angular velocity of a star's actual motion across our line of sight, on the celestial sphere.

Protostar. A contracting region of gas and dust that may lead to the formation of a new star.

Protostellar disk. The circular zone surrounding a nascent star, where planets may eventually form.

Pulsar. A supernova remnant that is a source of rapidly pulsating radio emissions with periods of a few seconds or less. These objects are manifestations of neutron stars having bipolar beams of energy that happen to periodically aim toward Earth. Such objects have diameters on the same order of magnitude as Earth.

Pulsating variable stars. Stars that vary in both luminosity and diameter over certain periods of time.

Quasar. These objects appear to have star like centers and are slightly variable in brightness, with periods of a few years or less. They are characterized by large redshifts, which indicate distances from Earth on the order of billions of light years. They are believed to be sites of distant supermassive black holes in the center of galaxies, where huge amounts of matter is accelerated and heated to extremely high temperature and luminosity as the swirling material orbits ever closer to the central black holes.

Radial velocity. The component of a star's space motion directly towards or away from the earth. A positive (+) radial velocity means the object is moving away from earth and negative (-) implies motion towards the earth. Radial velocities are determined by measurements of Doppler shifts in the spectra of celestial objects.

Radiation. The way electromagnetic energy is transmitted across space.

Radio astronomy. The branch of astronomy that deals most directly with observations of electromagnetic energy having wavelengths of one millimeter or longer emitted by celestial objects.

Radio telescope. A device used to collect and detect radio energy emitted by celestial objects.

Recurrent nova. A nova whose eruption has been observed on more than one occasion.

Red giant. A star in the late stages of its evolution, whose luminosity and color place it on the upper-right-hand portion of the H-R diagram, above the main sequence. Although red is the most intense color produced by these stars, the blend of all colors radiated by red giants causes these stars to appear yellow or yellow-orange to our eyes.

Redshift. A shift of absorption or emission lines towards the red end of an object's spectrum. Redshifts may be caused by either an object's motion away from the earth or by the effects of an intense gravitational field. Redshifts provide a means for estimating the distances of galaxies more than a few hundred million light years from Earth.

Reflecting telescope. A telescope that uses a concave mirror to collect light and to form images.

Reflection nebula. A cloud of interstellar dust visible because of its reflection of light from neighboring stars.

Refracting telescope. A type of telescope that uses a lens or set of lenses to collect light and form an image.

Resolution. The ability of a telescope or other instrument to detect the separations of double stars or to show fine details of an object.

Revolution. Cyclic motion of one object around another.

Right ascension. A numbering system for equatorial coordinates used to indicate the east-west position of an object on the celestial sphere. The coordinates of right ascension range from zero through 23 hours and back to zero, a span of 24 hours. Each hour of right ascension is divided into minutes and seconds. The minutes (m) and seconds (s) of right ascension are related to times and are not equal to the minutes (') and seconds (") of angular measure, which are used to subdivide degrees of declination. The great circle that indicates zero hours of right ascension passes through the celestial poles and the point of the Vernal Equinox on the celestial equator.

Rotation. Motion by which an object spins around an axis that passes through itself.

Royal Family of the Sky. Group of constellations – Andromeda, Perseus, Cepheus, and Cassiopeia, seen together during evenings in autumn.

RR Lyrae stars. A type of pulsating variable star that has a period of less than one day and is usually found in the Galactic halo or bulge. As is the case with Cepheid variables, the period-luminosity relation for RR Lyrae stars serves as an important tool for estimating distances in space. RR Lyrae stars are also known as cluster-type variables, since they are found in most globular clusters.

Schwarzschild radius. The radius at which a massive contracting object becomes a black hole. When this size is reached, light from the object is no longer able to escape the object's highly concentrated gravitational field.

Seeing conditions. The conditions in the atmosphere that affect our ability to see well-defined images through a telescope.

Semi-detached eclipsing binary. A double star system whose components are close and share a physical link of matter partially bridging the space between them. Furthermore, they orbit their common center of mass in a plane whose edge is directed straight toward Earth. As a result, each component star periodically partly or completely eclipses the other, as seen from our perspective.

Semi-regular variable star. Supergiant or giant stars with non-constant variability periods, usually in the range of one or two magnitudes over periods from several weeks to about half a dozen years.

Separation (of double stars). The angular separation between double stars.

Shock wave. A component of compressed gas that expands through a medium as a result of the passage of material through the material at a speed greater than that of sound.

Sidereal day. The time interval between two successive transits of the celestial meridian by the Vernal Equinox.

Sidereal time. The number of hours, minutes, and seconds since the Vernal Equinox last crossed the celestial meridian. This is equivalent to the right ascension of an object currently on the observer's meridian.

Singularity. A point at the center of a black hole where the known laws of physics apparently no longer hold true.

Sky Screen. A region of sky, extending from north through east to south, centered approximately 30° above the horizon, where stars featured in *The Star Guide* are presented on specific dates throughout the year at about 9:00 p.m. local time (10 p.m. when Daylight Saving is in effect).

Small circle. A circle on a spherical surface, defined by the intersection of the sphere with a plane that does not pass through its center.

Solar antapex. The apparent point on the celestial sphere that the Solar System is moving away from, relative to neighboring stars.

Solar apex. The point on the celestial sphere towards which the sun is apparently traveling, relative to other stars in its neighborhood.

Solar System. The gravitationally bound system that includes the Sun, planets, satellites, asteroids, comets, and meteoroids.

Solstitial colure. The meridian circle on the celestial sphere which passes through the celestial poles, zenith and the Sun's position on dates of the June and December equinoxes.

Space motion. The total velocity of a star as it moves through space, taking into account proper motion and radial velocity.

Speckle interferometry. A method for enhancing resolution of fine details in images of relatively bright celestial objects.

Spectral type. A classification of stars made on the basis of comparisons of various spectral lines. The' sequence of spectral types is designated 0, B, A, F, G, K, M, an order that represents decreasing surface temperatures.

Spectroscope. A device used to analyze light by separating components at various wavelengths.

Spectroscopic analysis. The study of a spectrum in order to learn about the source of its light.

Spectroscopic parallax. An estimate of stellar distance based on a comparison of a star's apparent magnitude with the absolute magnitude characteristic of its spectral and luminosity classes.
Spectrum. An array of colors and/or other manifestations of electromagnetic radiation.
Spiral arms. Regions of gas, dust, and young stars that curve outwards from the centers of spiral galaxies.
Spiral galaxy. A type of galaxy characterized by a flattened, rotating disk of stars, which contains a system of spiral arms.
Star. A hot sphere of gas that, by means of internal thermonuclear fusion reactions, shines with its own light.
Star cluster. A group of stars whose members are gravitationally bound together.
Stellar evolution. The changes which take place as stars form, mature, and expire.
Stellar nursery. A region where new stars are forming from dust and gas, in spiral arms of a galaxy's disk.
Stellar wind. A flow of material from a star's surface out into space.
Stromgren sphere. A region of heated and expanding hydrogen which surrounds an extremely hot and luminous young star or stars. Also called an H II region.
Subgiant. The first luminosity class (IV), attained by a star as it expands and evolves away from the main sequence to eventually become a giant.
Summer (June) solstice. The most northerly position of the Sun on the ecliptic. The Sun's arrival at this point, on or about June 21, marks the beginning of summer in the Northern Hemisphere.
Sun. The star nearest to Earth.
Supercluster. A group of many individual clusters of galaxies extending over a region of space having a diameter on the order of 200 million light years.
Supergiant. A star of exceptionally high luminosity and diameter, assigned to luminosity classes Ia or Ib.
Supernova. A cataclysmic explosion in which a star of high mass blows itself apart and, in the process, increases in luminosity up to a billion times. If the star's core survives, it contracts to become a neutron star or black hole.
Supernova remnant. A neutron star, a black hole, or an expanding nebula, formed from the material ejected into space by a supernova.

T Tauri stars. A type of irregular variable star, still in the process of pre-main sequence gravitational contraction.
Telescope. An instrument used to collect and magnify electromagnetic radiation from distant objects.
Transit. The passage of a celestial object across the celestial meridian.
Ultraviolet radiation. A form of electromagnetic radiation having wave-lengths just shorter than violet.
Universe. Entirety of everything known to exist.

Variable star. A star that varies in brightness due to systematic internal instabilities or periodic blocking of its light by an orbiting partner.
Vernal Equinox. The point at which the ecliptic intersects the celestial equator and which is the location of the sun on or about March 21. The Sun's arrival at the Vernal Equinox marks the beginning of spring in the Northern Hemisphere.
Very Large Array (VLA). A group of twenty-seven radio telescopes, each with a dish antenna having a diameter of 25 meters, located along three 21-kilometer-long tracks in New Mexico. When these telescopes are used in conjunction with each other, high-resolution radio images of celestial radio sources are obtained.
Visual double star. A double star whose components have sufficient angular separation so that they may be seen as double when viewed through a telescope. Such stars may be gravitationally bound partners, or may be unrelated in which case they are known as optical doubles. (think "optical illusion")

Wavelength. Distance between comparable points of an electromagnetic wave.
White dwarfs. The dense, collapsed remnants of stars near the end of their evolution.
Winter (December) solstice. The most southerly point on the ecliptic, reached by the Sun on or about December 21. This event marks the start of winter in the Northern Hemisphere.
Zenith. The point located 90 degrees from the horizon, directly over an observer's head.
Zenith Distance. Angular separation of a celestial object from the zenith.
Zero-age main sequence. (ZAMS) The reclining "S" shaped line on an H-R diagram occupied by stars which have just arrived on the main sequence, after having begun fusing hydrogen in their cores and ceased gravitational contraction.
Zero hour of right ascension. Defined by the semicircle on the celestial sphere passing through the Vernal Equinox as well as the celestial poles, this coordinate represents the starting line for measuring right ascension.
Zodiac. A circular region on the celestial sphere 16 degrees or arc wide centered on the ecliptic. The zodiac is where the Sun, Moon and planets may be seen.

Bibliography

Sky Observations

Allen, Richard Hinckley, *Star Names: Their Lore and Meaning.* New York: Dover Publications, 1963.
Burnham, Robert, Jr., *Burnham's Celestial Handbook.* 3 vols. New York: Dover Publications, 1978.
Chartrand, Mark R., III, *Skyguide.* Racine, WI: Golden Press, 1983.
Cox, Arthur N., Allen's Astrophysical Quantities 4th ed. New York: Springer-Verlag, 2000.
Dickinson, Terence, *NightWatch*. Richmond Hill: Firefly Books, 2006.
French, Sue, *Deep-Sky Wonders*. Richmond Hill: Firefly Books, 2011.
Gallant, Roy A., *The Constellations.* New York: Four Winds Press, 1979.
Kitchin, C. R., *Astrophysical Techniques, 5th ed.* Boca Raton: CRC Press, 2009.
Krupp, E. C., *Beyond the Blue Horizon: Myths and Legends of the Sun, Moon, Stars, and Planets*. Oxford: Oxford University Press, 1992.
Kunitzsch, Paul, *ArabischeSternnamen in Europe.* Wiesbaden, Germany: O. Harrossowitz, 1959.
Kunitzsch, Paul, and Smart, Tim, *A Dictionary of Modern Star Names: A Short Guide to 254 Star Names and Their Derivations*. Cambridge MA: Sky Publishing, 2006.
Loeb, Avi, *extraterrestrial, The First Sign of Intelligent Life Beyond Earth,* Boston: Houghton Mifflin Harcourt, 2021.
Martin, R. Newton, and Menzel, Donald Howard, *The Friendly Stars.* New York: Dover Publications, 1964.
Mayall, R. Newton, and Mayall, Margaret W., *Olcott's Field Book of the Skies.* 4th ed. New York: G. P. Putnam's Sons, 1954.
Mayall, R. Newton; Mayall, Margaret W.; and Wyckoff, J., *Sky Observer's Guide.* New York: Golden Press, 1977.
Muirden, James, *The Amateur Astronomer's Handbook.* 3d ed. New York: Harper & Row, 1982.
Mullaney, James, and McCall, Wallace, *The Finest Deep-Sky Objects.* Cambridge, MA: Sky Publishing, 1978.
O'Meara, Stephen James, *Deep-Sky Companions: The Messier Objects 2nd ed.* New York: Cambridge University Press, 2014. With Mario Motta, *Deep-Sky Companions: The Caldwell Objects*, *2nd ed.* New York: Cambridge University Press, 2016.
Ottewell, Guy, *Astronomical Calendar.* Annual. Greenville, SC: Furman University. *Astronomical Companion.* Greenville, SC: Furman University.
Pasachoff, Jay M., Tirion Wil, *Field Guide to the Stars and Planets.* New York: Peterson Field Guides/Houghton Mifflin, 2000.
Peltier, Leslie c., Starlight Nights: The Adventures of a Star-Gazer. Cambridge, MA: Sky Publishing, 1980.
Proctor, Percy M., Star Myths and Stories. New York: Exposition, 1972.
Ray, H., *The Stars: A New Way to See Them.* 3d ed. Boston: Houghton Mifflin, 1967.
Royal Astronomical Society of Canada, *Observer's Handbook.* Annual. Toronto: University of Toronto Press.
Sherrod, P. Clay, *A Complete Manual of Amateur Astronomy.* Englewood Cliffs, NJ: Prentice-Hall, 1981.
Staal, Julius, D. W., *New Patterns in the Sky: Myths and Legends of the Star.* McDonald & Woodward Publishing Company, 1988.
Strand, Kaj Aage, ed., *Basic Astronomical Data*. Chicago: University of Chicago Press, 1963.
U.S. Naval Observatory, *The Astronomical Almanac.* Annual. Washington, DC: U.S. Government Printing Office.

Star Atlases

Cambridge Double Star Atlas: James Mullaney and Wil Tirion (to mag. 7.5)
Men, Monsters and the Modern Universe: George Lovi and Wil Tirion
Millennium Star Atlas: Roger Sinnott, Michael Perryman (to mag. 11)
Norton's Star Atlas and Reference Handbook: Ian Ridpath (to mag. 6.5)
Pocket Sky Atlas*:* Roger Sinnott (stars to magnitude 7.5)
SkyAtlas 2000.0, second edition: Wil Tirion & Roger Sinnott (to mag. 8.5)

Various Aspects

Allday, Jonathan, *Quarks, Leptons and the Big Bang*: Third Edition. Boca Raton: CRC Press, 2017.
Ashbrook, Joseph, *The Astronomical Scrapbook: Skywatchers, Pioneers, and Seekers in Astronomy.* edited by Leif J. Robinson. Cambridge, MA: Sky Publishing Corp, 1985.

Asimov, Isaac. *The Universe: From Flat Earth to Quasar.* New York: Discus, Avon, 1976.

Ball, Nadine Butcher, Coyle, Harold P., Shapiro, Irvin I., eds. *Project Spica: A Teacher Resource to Enhance Astronomy Education*. Dubuque: Kendall/Hunt, 1994.

Bacon, Marges, *Ernest Flagg: Beaux-Arts Architect and Urban Reformer.* New York: The Architectural History Foundation; Cambridge, MA: The MIT Press, 1986.

Bedini, Silvio A., "The Scientific Instruments of The Lewis And Clark Expedition" (1984). *Great Plains Quarterly*.

Bok, Bart J., and Bok, Priscilla F., *The Milky Way.* 5th ed. Cambridge, MA: Harvard University Press, 1981.

Dreyer, J.L.E., *History of the Planetary Systems from Thales to Kepler*. Cambridge, UK: Cambridge University Press, 1906.

Einstein, Albert, *Relativity: The Special and General Theory.* New York: Crown, 1961.

Ferris, Timothy, *Galaxies.* San Francisco: Sierra Club Books, 1980. *The Red Limit.* New York: William Morrow, 1977.

Feynman, R. *The Character of Physical Law.* Cambridge, MA: MIT Press, 1994.

Gamow, *George, One, Two, Three, Infinity: Facts and Speculations of Science*. New York: Dover/Viking, 1961.

Greene, Brian R., The Elegant Universe. New York: W. W. Norton & Co., 2003.

Harari, Yuval Noah, *Sapiens: A Brief History of Humankind*. New York: HarperCollins, 2015.

Hawking, Stephen, and Leonard Mlodinow, *The Grand Design.* New York: Bantam Books, Random House, 2010. *A Brief History of Time*. New York: Bantam Books, 1988.

Hopkins, Jeanne, *Glossary of Astronomy and Astrophysics.* 2nd. ed. Chicago: University of Chicago Press, 1980.

Hutchins, Robert Maynard ed., *Great Books of the Western World, Ptolemy, Copernicus, Kepler.* Chicago: Britannica, 1952.

Kaku, Michio, *The Future of the Mind.* New York: Anchor Books, 2015. *Physics of the Future.* New York: Doubleday, 2011. *Physics of the Impossible.* New York: Anchor Books, 2009. *Parallel Worlds.* New York: Anchor Books, 2005. *Hyperspace*. New York: Doubleday, 1994.

King, Henry C., *The History of the Telescope.* New York: Dover Publications, 1979.

Kippenhahn, Rudolf, *100 Billion Suns: The Birth, Life, and Death of the Stars.* New York: Basic Books, 1983.

Koch, Kenneth, and Farrell, Kate, *Talking to the Sun: An Illustrated Anthology of Poems for Young People*. New York: The Metropolitan Museum of Art, Henry Holt and Company, 1985.

Malin, David, and Murdin, Paul. *Colours of the Stars.* Cambridge, UK: Cambridge University Press, 1984.

Ostriker, Jeremiah P., and Mitton, Simon, *Heart of Darkness.* Princeton: Princeton University Press, 2013.

Peebles, P.J.E., *Cosmology's Century: An Inside Hisory of Our Modern Understanding of the Universe.* Princeton: Princeton University Press, 2020

Peebles, P. James E., Page, Lyman A., and Partridge, R. Bruce., *Finding the Big Bang.* Cambridge, UK: Cambridge University Press, 2009.

Preston, Richard S. "The Accuracy of the Astronomical Observations of Lewis and Clark." *Proceedings of the American Philosophical Society*, vol. 144, no. 2, 2000, pp. 168–191

Sagan, Carl, *The Cosmic Connection: An Extraterrestrial Perspective.* New York: Anchor Press- Doubleday, 1980. *Cosmos.* New York: Random House, 1980. *Contact.* New York: Simon & Schuster, 1986.

Sobel, Dava, *Glass Universe, How The Ladies of the Harvard College Observatory Took The Measure of the Stars.* New York: Penguin Books, 2016. *Longitude*, Walker Publishing Company Inc, 1995.

Tyson, Neil deGrasse, *Astrophysics for People in a Hurry.* New York, London: W.W. Norton & Co., 2017. *Welcome to the Universe.* Princeton: Princeton University Press, 2016. *Space Chronicles.* New York: W.W. Norton & Co., 2012. *The Pluto Files.* New York: W.W. Norton & Co., 2009. *Death by Black Hole.* New York: W.W. Norton & Co., 2007.

Wallenquist, Ake, edited and translated from Swedish by Engelbrektson, Sune, *Dictionary of Astronomical Terms.* Garden City, New York: The Natural History Press, 1966.

Warner, Deborah Jean, *Alvan Clark and Sons: Artists in Optics.* Washington, DC: Smithsonian Institution Press, 1969. *The Sky Explored: Celestial Cartography 1500-1800.* New York: A. R. Liss, 1979.

Willard, Berton C., *Russell W. Porter: Arctic Explorer, Artist, Telescope Maker.* Freeport, ME: Bond Wheelwright, 1976.

Periodicals

I. Astronomy

Astronomy

Mercury Astronomical Society of the Pacific

Sky Calendar, Abrams Planetarium (monthly newsletter)

Sky and Telescope, American Astronomical Society

The Observer's Handbook, Royal Astronomical Society of Canada

II. Science in General, with Some Articles about Astronomy:

Discover

Natural History
New Scientist
Science News
Scientific American
Smithsonian Magazine

III. Professional Journals:

Astrophysics Data System – Smithsonian Astrophysical Observatory/NASA
APS News Regarding open source (arXiv, non-subscription) journal articles – American Physical Society
American Scientist
Annual Review of Astronomy and Astrophysics
Astronomy and Astrophysics
Astronomical Journal
Astrophysical Journal
Astrophysical Journal Letters
Icarus
Journal for the History of Astronomy
Monthly Notices of the Royal Astronomical Society
Nature
Philosophical Transactions of the Royal Society of London
Publications of the Astronomical Society of the Pacific

Index

S

T

U

V

W

Z

Made in United States
North Haven, CT
01 October 2023

42216446R00071